The AI Advantage for Product Managers

The AI Advantage Series · Book One

THE AI ADVANTAGE SERIES

The AI Advantage for Product Managers

Book Two: AI Product Strategy for Senior Leaders (forthcoming)

The AI Advantage
for Product Managers

*The Complete Playbook for Leading, Building, and Winning
in the Age of Intelligent Products*

Mario Urbina

The AI Advantage Series · Book One

This book is dedicated to my parents, my brothers and sisters for always being there when I need, their honesty, and the particular way to push me to be a better professional and better person. You shaped my career before I knew I had one and guided me through the hard times without ever making me feel lost.

To my friends, for always motivating me to keep going.

CONTENTS

FOREWORD

There is a book that sits on the desk of nearly every product manager I have ever hired. Some copies are dog-eared from years of use. Some look brand new, pulled out and reread before a difficult strategic conversation. A few have sticky notes on every other page. The book is Inspired by Marty Cagan, and for the better part of two decades it has been the closest thing the product management profession has had to a shared foundation, a common language for what it means to build products people love, with empowered teams, and with the rigor that separates genuine product thinking from the appearance of it.

I mention Inspired not to invoke it as an authority, but to mark a moment. Because for the first time in a long time, I believe we are at an inflection point significant enough to require a new foundation. Not a replacement for what Cagan gave us; the principles of empowered product teams, continuous discovery, and outcome-based thinking remain as true as they ever were. But a serious extension of that foundation into a world that looks fundamentally different from the one those principles were built for.

The world where AI is a feature you add to a product is already behind us. The world where AI is the product, where intelligent agents take consequential actions on behalf of users, where the design surface includes failure modes that traditional product thinking was never built to handle, where the speed of learning is not a nice-to-have but the primary competitive variable; that world is the one product managers are waking up to every morning right now. And most of us, if we are honest with ourselves, are still using the old maps.

I have spent the last three years working at the intersection of AI capability and product strategy, and the gap I keep seeing is not a technical gap. The engineers on most product teams are remarkably capable. The models are genuinely impressive. The gap is in product thinking, in the frameworks, the judgment, and the disciplined practice that turns AI capability into products that users trust, that businesses can measure, and that compound in value over time rather than flattening out after the demo.

That gap is what this book closes.

What you will find in these pages is not a survey of AI tools or a breathless account of what large language models can theoretically do. It is a practitioner's playbook, the kind that gets written by someone who has been in the room when the agent takes the wrong action, when the eval framework catches the failure that would have reached a hundred thousand users, when the launch communication overpromises and the trust account starts draining before the product has had a chance to prove itself.

The author has done something genuinely difficult: taken a fast-moving, technically complex, strategically consequential domain and made it legible; not by simplifying it into uselessness, but by finding the level of abstraction at which product managers can make better decisions. The AI literacy chapter is the clearest explanation of what PMs actually need to know about LLMs, agents, and evals that I have encountered anywhere. The chapter on designing for agentic AI names failure modes I have lived through and wished I had a framework for at the time. The career section is the honest conversation the industry has been having in private but not yet committing to in print.

If you are a mid-level PM wondering whether AI is going to make your role redundant or elevate it, this book has your answer, and it is the right one. If you are a senior leader trying to build an AI product organization that is genuinely capable rather than aspirationally positioned, Part Two and Part Four were written for you. If you are somewhere in between, trying to figure out which side of the AI PM divide you want to stand on, the introduction will orient you in about seven pages and you will not need to wonder anymore.

The book that helped a generation of product managers understand what good product work looked like in the software era was written at exactly the right moment. I believe this book is being written at exactly the right moment too. The profession needs it. The practitioners navigating this transition deserve it.

Read it once to get the map. Read it again when you are building something hard. Keep it close to the work.

Mario Urbina, 2026

INTRODUCTION

The Split Has Already Happened

And the distance between the two sides is growing by the day.

> *Two product managers walk into the same company on the same day. Same title. Same job description. Five years from now, one of them is earning $245,000. The other is earning $123,000. The difference isn't seniority, pedigree, or luck. It's one decision (made right now) about which side of the AI divide to stand on.*

I'm going to be direct with you from the first page, because I think you deserve that: **the product management profession is splitting in two,** and it is happening faster than most people in the industry want to admit.

On one side, you have product managers who are learning to lead, build, and ship in a world where artificial intelligence isn't a feature, it's the foundation. They understand how intelligent systems work well enough to make better decisions. They know how to design for AI agents, measure AI outcomes, and build the organizational muscle to move at the speed the market now demands. They are not necessarily technical. But they are fluent.

On the other side, you have product managers who are still treating AI as a tooling upgrade. Who thinks prompt engineering is something engineers do. Who are adding "AI experience" to their resumes because they approved a feature that called an API. Who will walk into their next performance review not knowing that their company is quietly looking for someone who can do what they haven't learned yet.

This book is for the first group or for anyone who wants to get there.

The Numbers Tell the Story

In 2026, the salary data became impossible to ignore. Research tracking product manager compensation in the United States found a gap so significant it stopped conversations in rooms full of senior leaders:

$245,000 - average US compensation for AI-focused Product Managers

$123,000 - average US compensation for traditional Product Managers

$122,000 - the gap, representing an entirely different career ceiling

That gap is not a bonus. It is not a specialty premium that will normalize over time. It represents **a different career track altogether**, one with a different ceiling, different leverage, and a fundamentally different relationship with the companies being built around us.

But here is what the salary data doesn't capture: the companies paying those numbers are not just looking for PMs who know what an LLM is. They are looking for PMs who can lead AI product strategy without needing their engineering teams to translate everything. Who can walk into a room with a CPO and speak to business outcomes, not just model capabilities. Who can ship intelligent products, and then measure, iterate, and improve them in a market where the rules keep changing.

That combination of skills, strategic, practical, and deeply grounded in how AI actually works, is what this book is designed to give you.

· · ·

Why I Wrote This

I have been building products at the intersection of technology and business for long enough to have lived through several "these changes everything" moments. Most of them changed some things. A few of them changed a lot of things. Generative AI, and now, agentic AI, is in a category by itself.

What I kept running into, as AI moved from experiment to expectation, was a specific kind of frustration. Product managers who were sharp, experienced, and genuinely talented were struggling, not because they couldn't learn, but because the resources available to them were either too technical, too theoretical, or already out of date. Books written in 2022 missed the generative AI wave entirely. Books written in 2024 caught the wave but missed how quickly agentic AI would change the product design surface. And courses that promised to make you an "AI PM" in a weekend mostly taught you to use tools, not to think differently.

I wanted to write the book I wish I'd had one that is honest about what the AI era actually demands, practical about how to meet that demand, and respectful of the fact that you already know how to do your job. You don't need to be re-taught in product management. You need a clear-eyed guide to what has changed, what that means for how you work, and what you need to add to your repertoire to stay ahead.

A WORD ON TECHNICAL BACKGROUND

You do not need an engineering degree to get full value from this book. You do not need to know how to train a model, write a prompt in code, or architect a data pipeline. What you do need is a willingness to develop genuine fluency (not superficial familiarity) with the concepts that shape how AI products get built and how they succeed or fail.

There is a difference between knowing what RAG stands for and understanding why it matters for your product decisions. This book aims for the second. Every technical concept introduced here is introduced in service of a practical outcome, a better decision, a sharper strategy, a more effective conversation with your team.

. . .

What This Book Will Do For You

By the time you finish the last page, you will be able to do three things that most product managers (even experienced ones) currently cannot do well:

1. **Lead AI product strategy at your company.**

    You'll have the frameworks, the vocabulary, and the decision-making tools to set and defend an AI product strategy, from adaptive planning and accelerator squad design to agentic AI orchestration and the learning systems that make strategy durable.

2. **Build and ship intelligent products that actually work.**

    You'll know how to scope AI features correctly, collaborate effectively with ML engineers and data scientists, design for agentic experiences, measure what actually matters, and launch AI products that earn user trust rather than destroy it.

3. **Win the career on the right side of the divide.**

    You'll understand the four AI PM specializations, what companies like Google, Amazon, Microsoft, and Meta are actually hiring for, how to build a portfolio that signals genuine capability, and how to negotiate for the compensation that reflects it.

Those three outcomes are not abstract. Each one has a dedicated section in this book, grounded in real practitioner experience, current industry data, and case studies from the companies that are doing this well, and a few that aren't.

. . .

How to Use This Book

This book is structured in four parts that build on each other, but they are also designed to be useful individually, depending on where you are right now.

New to AI PM

Start at Chapter 1 and read straight through. Parts I and III will give you the foundation and tactical tools; Part IV will give you the career roadmap.

Mid-level PM in transition

You probably know the basics. Start with Chapter 3 to orient yourself in the landscape, then go directly to Parts II and III for strategy and execution depth.

Senior Leader or CPO

Jump to Part II for strategy and Chapter 11 for team building. Come back to Part I when you need to help your team develop their foundation.

Using the AI Advantage Training App

Each chapter maps to a module in the companion training app. The app is where you practice, this is where you build the mental model first. Access the training at **booksbymario.com/learn**

A note on the practitioner stories throughout this book: they are real. The names and some identifying details have been changed, but the decisions, the mistakes, and the lessons are drawn from actual product work. I believe strongly that the most useful thing a book like this can do is not give you more frameworks to memorize, it is to show you how a practitioner actually thinks in the moment, when the stakes are real and the answer isn't obvious.

. . .

One More Thing

The split I described at the beginning of this introduction, the one between the product managers earning $245,000 and the ones earning $123,000, is not inevitable. It is not a matter of raw talent. It is not determined by the school you attended, the company you started at, or how technical your background is.

It is determined by a decision. A decision to develop genuine capability in this new era of products, rather than the surface-level familiarity that too many people are settling for.

You picked up this book. That's the first part of the decision. Let's make sure the rest of it counts.

— *Mario Urbina, May 2026*

PART ONE: UNDERSTAND THE SHIFT

CHAPTER ONE

The New Rules of Product Management

What changed, what it means, and why most product teams are still playing by the old rulebook.

In the spring of 2025, Slack's CPO stood in front of a room of senior product leaders and said something that made half of them uncomfortable: "The roadmap is the wrong artifact." Not outdated. Not imperfect. Wrong. He wasn't talking about a better template or a more collaborative process. He was talking about a fundamentally different way of thinking about how product work gets done.

That moment is a useful place to start this book, because it captures something important about where product management is right now. The old rulebook is not merely being updated. In several critical areas, it is being replaced. And the product managers who are thriving are not the ones who found a better version of the old rules. They are the ones who understood that the game itself had changed.

This chapter is about those changes. Not as a survey of trends, but as a practical reset of the mental model you will need to lead, build, and ship in an AI-driven market. Some of what follows will confirm things you have already sensed. Some of it will challenge assumptions that have served you well until recently. All of it is grounded in what the most effective product leaders are actually doing right now.

. . .

Rule One: Roadmaps Are Hypotheses, Not Plans

For most of the past two decades, the product roadmap was the central artifact of the product management role. It was how you communicated strategy to stakeholders, aligned engineering teams, managed expectations with sales, and demonstrated progress to leadership. A good roadmap, in that world, was a sign of a good product manager.

That world is gone. Not because roadmaps were always wrong (they weren't), but because the conditions that made them useful have changed. When AI allows any competitor to clone your product's experience, workflows, and messaging in a matter of weeks, a six-month roadmap is not a strategy. It is a document that will be partially obsolete before it is finished.

What replaces it is not chaos. It is a combination of **clear product principles, outcome-based objectives, and rapid AI-assisted prototyping.** Slack's approach is instructive. Rather than committing to a detailed feature plan, the team operates with a set of product principles that guide day-to-day decisions, a small number of measurable outcomes they are accountable for, and cross-functional squads (sometimes as small as one designer and one engineer) who use AI to prototype constantly, learn quickly, and discard dead ends without hesitation.

The shift here is subtle but consequential. Planning for outcomes rather than outputs is not a new idea in product management. What is new is the speed at which this cycle now operates, and the role AI plays in compressing the distance between hypothesis and evidence. A prototype that once took two weeks to build can now be built in an afternoon. That changes what it means to make a product bet.

WHAT THIS MEANS IN PRACTICE

If your roadmap is the primary artifact you use to communicate product strategy, start shifting now. Define three to five product principles that any teammate could use to make a decision without asking you. Anchor your planning to outcomes (what changes for the user or the business) rather than outputs (what ships). Treat every item on your roadmap as a hypothesis that needs to be validated, not a commitment that needs to be delivered.

. . .

Rule Two: Learning Speed Is the New Competitive Moat

Andrey Khusid, the founder and CEO of Miro, was asked at a product conference what he considered the most important competitive advantage a company could have in 2026. His answer surprised people who were expecting something about technology or talent.

He said it was the speed of learning. Specifically: how fast an organization can recognize a signal in the market, separate it from noise, and act on it. Not how fast it can ship. Not how much data it has. How fast it can update its beliefs and change course.

This matters deeply for product managers, because it reframes what the job actually is. In the old model, the PM's job was to define the right features and shepherd them to delivery. In the new model, the PM's job is to run a learning operation. Discovery is not a phase that happens before development. It is a continuous practice that runs alongside it, powered by AI tools that make experimentation faster and cheaper than it has ever been.

The organizations that are winning have baked learning into their operating rhythm as a first-class activity. They do not treat retrospectives as a box to check. They do not wait for quarterly business reviews to update their assumptions. They have **rituals, tooling, and incentive structures** that reward learning as much as they reward shipping. We will go deep on how to build this kind of learning operating system in Chapter 6. For now, the important shift is conceptual: if you are measuring your team's success primarily by what shipped, you are measuring the wrong thing.

The strongest product organizations treat learning as infrastructure, not as an activity that happens when there is time for it. Build it into your team's rituals, tools, and incentives before you need it.

. . .

Rule Three: The Three Pressures Every PM Faces in 2026

Beyond the specific shifts in how roadmaps work and how learning gets structured, there is a set of broader pressures that nearly every product manager is navigating right now. Understanding them as a system (rather than as separate problems) changes how you respond to each one.

1. From buzzwords to defensible business outcomes.

AI has generated more product theater than almost any technology in recent memory. Features shipped to satisfy investor narratives. Roadmap items labeled as "AI-powered" because they called a third-party API. Demos that looked impressive and products that didn't work. The market has started to correct for this, and the correction is visible in how enterprise buyers evaluate AI products, how boards ask about AI ROI, and how the most respected product leaders talk about what they are actually building. The pressure on product managers is to move from AI as a feature to AI as a source of genuine, measurable business value. That requires a different kind of rigor in how you scope AI work, how you measure it, and how you talk about it.

2. From fragmented AI features to coherent AI systems.

Most product organizations that have been experimenting with AI for the past two years are sitting on a collection of disconnected AI features, each built by a different team, each with its own data assumptions, each measured differently. The next phase of AI product work is not adding more features. It is consolidating what exists into a coherent architecture that can actually scale, that users can understand and trust, and that the business can build on. This is a harder problem than it sounds, and it is the problem that separates the AI PMs who can lead at a systems level from those who can only manage individual features.

3. From rapid deployment to responsible governance.

As AI products move from internal tools to customer-facing experiences, the governance question becomes unavoidable. Not just as a legal or compliance concern, but as a product design problem. How do you build AI products that users can trust? How do you create explainability without destroying the user experience? How do you catch failure modes before they cause harm at scale? These are product management questions, and the PMs who are answering them well are the ones who treat governance as a design constraint from the beginning, not as a review step at the end.

These three pressures are not sequential. They are simultaneous. And the product managers who are navigating all three at once are developing a kind of capacity that is genuinely difficult to replicate quickly. That is the nature of a real competitive advantage.

. . .

Rule Four: AI Does Not Replace the PM. It Raises the Bar.

There is a version of the AI and product management conversation that you have probably heard more than once in the past two years: the one about whether AI will replace product managers. It is worth addressing directly, not because it is the most important question, but because the anxiety it generates can cloud clearer thinking about what is actually happening.

AI is not replacing product managers. But it is doing something that has a similar effect on the careers of PMs who do not adapt, it is automating the parts of the job that were never really the job. The status update decks. The requirements documents assembled from meeting notes. The first draft of the PRD. The competitive analysis that used to take a week. These tasks are not disappearing from the product organization. They are being absorbed by tools, freeing up time that should be redirected to the work that AI cannot do.

What AI cannot do is **determine the why or manage the who.** It cannot understand a company's mission and competitive nuances well enough to own the product vision. It cannot sit in a difficult conversation between engineering and sales and find the frame that moves both teams forward. It cannot read between the lines in a user interview, sense the hesitation in a customer's voice, or build the trust with a cross-functional partner that makes hard decisions easier. These are the things that great product managers do, and they are the things that AI makes more important, not less.

The PMs who are thriving in this environment are not the ones who have learned to use the most AI tools. They are the ones who have used AI to reclaim time for the highest-leverage version of their job: strategic thinking, deep customer empathy, and the organizational influence that turns good ideas into shipped products.

IN PRACTICE: THE DECISION THAT CHANGED HOW I THINK ABOUT AI

Early in my time building AI-assisted features, I made a mistake that I have seen many product managers make since. We had a clear user problem, a capable engineering team, and a genuinely interesting AI approach to solving it. We moved fast. We shipped.

The feature worked, technically. The AI did what we built it to do. But we had not spent enough time thinking about what the experience felt like when the AI got it wrong (which it did, more than we expected in the first weeks). We had not designed the failure states. We had not given users a clear way to understand why the AI had made a particular suggestion, or a graceful path to override it. The trust we needed to build with users in the first few weeks of the launch was damaged by problems we could have anticipated if we had asked different questions earlier.

That experience taught me something I have returned to many times since: with AI products, the quality of your thinking before you build matters more than it does with traditional software. The blast radius of a bad design decision is larger, faster, and harder to recover from. The new rules of product management are not just about going faster. They are about thinking more carefully, earlier, about the right things.

What the New Rules Add Up To

Taken together, the shifts described in this chapter point toward a version of the product management role that is more strategic, more systems-oriented, and more deeply connected to business outcomes than the version that most of us were trained for.

Roadmaps become hypotheses. Learning becomes infrastructure. The three pressures (outcomes, coherence, governance) become a permanent feature of the landscape rather than a phase to manage. And AI becomes a force multiplier for the work that only humans can do, which means the quality of that human work matters more than ever.

This is not a harder version of the old job. It is a different job. And the good news is that the skills it requires are learnable, the frameworks for developing them are clear, and the product managers who invest in them now are building an advantage that will compound over the next five to ten years.

The next chapter gives you the foundation you need to operate confidently in this environment: the technical literacy that lets you make better product decisions without becoming an engineer. We will cover what you actually need to know about LLMs, agents, RAG, and evals, and (just as importantly) what you do not need to know and why.

— · —

Continue to Chapter 2: AI Literacy Without the Engineering Degree

CHAPTER TWO

AI Literacy Without the Engineering Degree

What you actually need to know, what you do not, and why the difference matters more than most people think.

A few years ago, a senior product manager at a well-known enterprise software company asked her engineering lead to explain how the company's new AI recommendation feature worked. He spent twenty minutes walking her through neural network architecture, loss functions, and gradient descent. She nodded throughout. She understood almost none of it. She went back to her desk, made a product decision based on incomplete information, and the feature shipped with a fundamental design flaw that took six months to fix. She did not need to understand gradient descent. She needed to understand one thing: that the model's recommendations would degrade significantly when the training data did not reflect the user's current context. That one insight would have changed her decision. The other nineteen minutes were noise.

The problem with most AI literacy resources for product managers is that they start from the wrong question. They ask: what does a product manager need to know about AI? And they answer that question by working through the technology from the ground up, covering concepts in the order an engineer would learn them rather than in the order a PM needs them to make better product decisions.

This chapter starts from a different question: what AI concepts actually change the decisions you make? The answer is a much smaller and more useful set of ideas than most AI literacy courses would have you believe. You do not need to be able to train a model. You do not need to understand backpropagation. You do not need to know the difference between a transformer and a recurrent neural network unless you are building at the infrastructure layer (and if you are, this is not the chapter for you).

What you do need is the ability to **ask the right questions, recognize when something is technically feasible, understand the failure modes that will affect your users, and make informed trade-off decisions without having to outsource your judgment to the engineering team.** That is what this chapter gives you.

. . .

The Concepts That Actually Matter: LLMs

Large language models are the foundation of most AI products being built today. If you are building anything that involves generating text, summarizing content, answering questions, writing code, or having a conversation with a user, you are almost certainly building on top of an LLM. Here is what you need to understand about them as a product manager.

An LLM is, at its core, a prediction engine. It has been trained on enormous amounts of text and has learned, through that training, to predict what word (or token) is most likely to come next given everything that came before. That sounds simple. The implications for product design are not.

The first implication is that LLMs do not know things the way a database knows things. A database either has a record or it does not. An LLM generates a plausible-sounding response based on patterns in its training data. This means it can be confidently wrong. It can produce an answer that sounds authoritative and is factually incorrect. As a product manager, this is not a bug you can fix by upgrading the model. It is a property of how these systems work, and your product needs to be designed around it.

The second implication is that LLM behavior is **probabilistic, not deterministic.** Ask the same question twice and you may get two different answers. This is useful when you want creative output and deeply problematic when you need consistent, auditable decisions. Understanding where on this spectrum your product sits is one of the most important product decisions you will make early in the design process.

The third implication is about context windows. An LLM can only work with what is in its context at the time of inference (the moment it generates a response). It does not have persistent memory between conversations unless you build that memory yourself. It cannot access real-time information unless you provide it. These constraints shape what your product can and cannot do, and they are the reason that retrieval-augmented generation (covered shortly) became so important so quickly.

THE THREE LLM QUESTIONS EVERY PM SHOULD BE ABLE TO ANSWER

1. What happens when the model is confidently wrong? How does your product handle it, and how does the user recover?
2. Where does your product need deterministic output, and where can it tolerate probabilistic variation? Have you designed differently for each?
3. What does your model know, and what does it not know? Where are the gaps between its training data and the reality your users live in?

. . .

Agents: What They Are and Why They Change Everything

For the first few years of the modern AI era, most AI products followed a simple pattern: a user puts in a prompt, the model generates a response, the user reads it. That pattern is already starting to feel dated. The products being built now (and the ones your company will be competing with in the next eighteen months) are increasingly built on AI agents.

An AI agent is a system that can **take actions, not just generate responses.** Where a basic LLM interaction ends with text on a screen, an agent can use that text to trigger a tool, query a database, send an email, update a record, make a decision, and then decide what to do next based on the result. It operates in a loop: perceive, reason, act, observe the outcome, repeat.

The practical difference for product design is significant. When you are building with a basic LLM, the main design surface is the prompt and the response. When you are building with agents, the design surface expands to include the tools the agent can use, the guardrails that constrain its behavior, the decision points where a human needs to review or approve, and the failure modes that occur when the agent takes an action based on incorrect information or misunderstood intent.

Multi-agent systems take this further. Rather than one agent doing everything, a multi-agent system assigns different agents to different specialized tasks and coordinates them through an orchestration layer. Walmart's implementation is a useful reference point: one system translates product content across twenty-two languages, another manages supplier onboarding, another handles inventory forecasting. Each agent is specialized. The orchestration layer routes work, monitors for anomalies, and keeps the system learning. The product management team's job shifted from managing features to managing a fleet.

The move from LLM features to agentic products is not an incremental upgrade. It is a different design problem. If your team is building agents and you are still thinking about prompts and responses, you are designing for the wrong surface.

. . .

RAG: Giving Your AI What It Does Not Know

One of the most common limitations PMs encounter when building AI products is the gap between what the model knows and what the user needs it to know. A general-purpose LLM trained on public data does not know your company's internal policies. It does not know your product's latest release notes. It does not know what happened in the news this morning. Retrieval-augmented generation (RAG) is the most widely adopted solution to this problem, and understanding it will make you a significantly more effective partner to your engineering team.

The concept is more intuitive than the name suggests. Rather than asking the model to generate a response from memory alone, a RAG system first retrieves relevant documents or data from an external source, then passes that retrieved content to the model along with the user's question. The model generates its response based on both its training and the retrieved context. Think of it as giving the model a relevant briefing document before it answers.

For product managers, RAG creates a set of design decisions that are genuinely consequential. The quality of a RAG system is heavily dependent on **the quality and freshness of what gets retrieved.** If the retrieval step surfaces outdated, irrelevant, or incomplete documents, the model's response will reflect that. Garbage in, confident garbage out. This means that maintaining the knowledge base your RAG system draws from is not an engineering task to be set and forgotten. It is an ongoing product responsibility.

WHAT RAG MEANS FOR YOUR PRODUCT ROADMAP

If your product uses RAG (or is planning to), add knowledge base quality to your definition of product health. Track how often retrieved content is outdated, irrelevant, or missing. Build a process for updating and curating the knowledge base as your product and company evolve. This is less glamorous than building new AI features, and it is often the difference between an AI product that users trust and one they quietly stop using.

. . .

Evals: How You Know If Your AI Product Is Working

In traditional software, testing is relatively straightforward. You define expected behavior, you write tests, you run them, they pass or they fail. With AI products, this breaks down quickly. When the output is generated text, a recommendation, or an agent's decision, there is often no single correct answer. Two responses can both be reasonable. Two responses can both be problematic in different ways. The question is not whether the output is correct. It is whether the output is good enough, by the standards that matter for your product.

This is what evals are for. Evaluations are the testing framework for AI systems, and as a product manager, you need to understand them at a conceptual level because they directly affect what you can and cannot promise to your users and your stakeholders.

Deterministic evals

These measure things that have clear right and wrong answers: did the model follow the format? Did it include the required fields? Did it stay within the character limit? These are the easy ones, and they should be automated.

Probabilistic evals

These measure things that require judgment: is this response helpful? Is this summary accurate? Is this recommendation appropriate for this user context? These are harder to automate and often require human raters or, increasingly, another LLM acting as a judge (the "LLM-as-judge" pattern).

User-signal evals

These measure real-world behavior: do users accept the AI's suggestions? Do they override them? Do they abandon the product after an AI error? These are often the most valuable signals and the ones most often missing from early-stage AI product measurement frameworks.

The practical implication for product managers is this: before you ship an AI feature, you should be able to answer three questions. What does good look like for this feature, specifically enough that you could measure it? What does failure look like, and how quickly will you know if it is happening? And who is responsible for monitoring this after launch? If you cannot answer those three questions, the feature is not ready to ship.

. . .

The AI Literacy Checklist

The goal of this chapter was not to make you an AI engineer. It was to give you enough conceptual grounding that your product decisions are informed by how these systems actually work, rather than by assumptions that are common among non-technical PMs and frequently wrong.

Here is a practical checklist. For any AI feature or product you are responsible for, you should be able to answer these ten questions without deferring entirely to engineering:

1. What model or system is powering this feature?

You do not need to know the architecture, but you should know whether it is a third-party API, a fine-tuned model, or something built in-house, and what that means for cost, latency, and control.

2. What does the model know, and where are its gaps?

What is the training data? What is the knowledge cutoff? Where will it be confidently wrong?

3. Where on the deterministic-to-probabilistic spectrum does this feature sit?

And have you designed the user experience to match? Probabilistic outputs need different UX treatment than deterministic ones.

4. How does the product behave when the AI is wrong?

What does the user see? Can they recover gracefully? Is there a way to override or correct the AI's output?

5. Is RAG being used, and if so, what is the quality of the retrieval layer?

How current is the knowledge base? Who owns it? How often is it updated?

6. Are we using agents, and if so, what can they do?

What tools do they have access to? What are the guardrails? What happens when they take a wrong action?

7. How are we evaluating quality before and after launch?

What are our deterministic evals? What are our probabilistic evals? How are we capturing user-signal feedback?

8. What does success look like, and how will we measure it?

Not just technical performance metrics, but business outcomes. What changes for the user or the company if this feature is working?

9. What is the cost model?

AI inference can be expensive. Do you understand the cost per query, the cost at scale, and how cost scales with usage?

10. What are the trust and transparency requirements?

Do users need to know they are interacting with AI? Do they need to understand how a recommendation was made? What are the regulatory considerations in your market?

IN PRACTICE: THE CONVERSATION THAT CHANGED HOW I WORK WITH ENGINEERS

Early in my experience building AI products, I sat in a lot of meetings where I nodded at things I did not understand and then made decisions based on my best guess about what the technical constraints actually were. I was not unintelligent. I was under-equipped. And the decisions I made reflected that.

The shift happened when I started asking a different kind of question in technical reviews. Not "how does this work" (which produces explanations calibrated for engineers) but "what does this mean for the user when it fails" (which produces information I could actually use).

That question unlocked something in how my engineering teams talked to me. They stopped assuming I wanted the technical explanation and started giving me the product-relevant framing instead. The meetings got shorter. The decisions got better. The AI literacy checklist in this chapter is a formalization of those questions. They are not designed to make you sound technical. They are designed to get you the information you need to do your job well.

. . .

What Comes Next

With the conceptual grounding from this chapter, you are equipped to orient yourself in the AI product landscape, which is exactly what Chapter 3 covers. You will learn how to map the full range of AI products from AI-enhanced to AI-native to agentic, understand where your product sits today, and think clearly about where it needs to go to remain competitive. You will also encounter a concept that no other AI product management book has addressed directly: the shift toward AI-mediated discovery, where products are increasingly being evaluated by AI systems before a human ever interacts with them. Understanding this shift will change how you think about product explainability, go-to-market strategy, and what it means to build a product that can be found, understood, and trusted in an AI-first world.

— · —

Continue to Chapter 3: The AI Product Landscape

CHAPTER THREE

The AI Product Landscape: Where You Play

How to orient yourself in the AI product universe, identify where your product sits today, and understand the forces that will determine where it needs to go.

In 2023, adding an AI chatbot to your product made you an early mover. In 2024, it made you a fast follower. In 2025, it made you table stakes. In 2026, it can make you look like you do not understand your own market. The question is no longer whether your product has AI in it. The question is whether the AI in your product is doing something that matters, in a way that is genuinely difficult to replicate, in a market that is evaluating your product through an entirely new set of lenses.

Most product managers think about their competitive landscape in terms of features: what does the competitor have that we do not, what do we have that they lack, where are the gaps we can close, and where are the gaps we can open. That framing still matters. But in an AI-driven market, it is insufficient because it misses the more important question: where in the AI product landscape are you actually positioned, and what does that position imply about your defensibility, your roadmap priorities, and the way users and AI systems will evaluate your product?

This chapter gives you a map. Not a definitive taxonomy (the landscape is moving too fast for that) but a set of orienting frameworks that will help you answer those questions with more clarity and confidence than most of your competitors can bring to the same conversation.

. . .

The Three Tiers of AI Products

The most useful way to orient yourself in the AI product landscape is not by technology stack or by industry vertical, but by the role AI plays in the core value your product delivers. That lens produces three distinct tiers, each with different competitive dynamics, different design challenges, and different implications for how you should be spending your time.

Tier 1 AI-Enhanced Products

Traditional products where AI has been added to improve specific features or workflows without fundamentally changing the product's core value proposition. The product works without the AI. The AI makes it better.

Examples: *A CRM with AI-generated email drafts. A document editor with AI summarization. A project management tool with AI-powered task prioritization.*

PM implication: The competitive risk here is commoditization. If your AI enhancement is powered by a third-party API that any competitor can also call, you have not built a moat. You have built a feature. The PM's job at this tier is to identify which AI enhancements create genuine behavior change in users (they adopt new workflows; they get meaningfully better outcomes) versus which ones are demos that look impressive and get ignored after the first week.

Tier 2 AI-Native Products

Products where AI is central to the core value proposition. The product could not exist, or would be radically diminished, without it. The AI is not a feature. It is the product.

Examples: *A legal research tool that synthesizes case law. A medical imaging platform that flags anomalies for radiologist review. A code generation tool that writes production-ready software from natural language descriptions.*

PM implication: The competitive dynamics here center on data, trust, and specialization. General-purpose AI cannot compete with a well-tuned, domain-specific model trained on high-quality proprietary data. The PM's job at this tier is to protect and expand the data advantage, build the evaluation infrastructure that proves quality, and earn the user trust that makes the product defensible.

Tier 3 Agentic Products

Products built on AI agents that take actions, make decisions, and operate across systems on behalf of users. The product is not just generating responses. It is doing work: autonomous, multi-step, consequential work.

Examples: *An AI that manages your calendar, email triage, and meeting preparation end-to-end. A supply chain optimization system that reroutes shipments in real time. A software development agent that takes a feature request from specification to pull request.*

PM implication: This is the frontier, and it is where the highest-value, highest-complexity product management challenges now live. The PM's job at this tier is to design for trust and transparency in systems that are making real decisions, build the human-in-the-loop architecture that keeps humans appropriately in control, and create the measurement infrastructure to know when the agents are doing their jobs well and when they are quietly creating problems. Chapter 5 goes deep on all of this.

Most product organizations are operating across more than one of these tiers simultaneously, which creates a coordination challenge that is easy to underestimate. The skills, design patterns, and measurement approaches that work at Tier 1 are not the same ones that work at Tier 3. A product team that treats all AI product work as the same kind of work, will consistently under-invest in the capabilities that matter most at each tier.

> **A DIAGNOSTIC QUESTION FOR YOUR PRODUCT**
> If you turned off all the AI in your product tomorrow, what would users lose? If the answer is a convenience feature they enjoy but do not depend on, you are at Tier 1. If the answer is the core reason they use your product, you are at Tier 2. If the answer is work that would otherwise not get done at all, you are approaching Tier 3. Knowing your tier tells you where to focus, what to measure, and what defensibility actually requires.

. . .

The Invisible Evaluation Layer: AI-Mediated Discovery

Here is a shift in the competitive landscape that most product managers are not yet designing for, even though it is already affecting their products: the way users discover, evaluate, and choose products is changing because AI systems are increasingly mediating those decisions.

When a user asks ChatGPT, Perplexity, Claude, or any AI assistant to recommend a project management tool, a CRM, a design platform, or a legal research service, that AI is making judgments about your product before any human ever interacts with it. It is synthesizing information from across the web, including your documentation, your marketing copy, your user reviews, and your API specifications, to form an assessment of what your product does, who it is for, and whether it is a good fit for the user's needs.

This creates a new competitive surface that most product teams are not managing: **product explainability.** Not explainability in the AI ethics sense (though that matters too), but explainability in the sense of how clearly your product communicates its purpose, value, behavior, and limits to AI systems that are evaluating it on behalf of potential users.

The products that perform well in AI-mediated discovery are not necessarily the best products. They are the products that are best understood by AI systems. And the gap between those two things is significant, actionable, and almost entirely unaddressed by the current generation of product teams.

. . .

What Product Explainability Actually Means

Explainability as a product surface area has four components, each of which you can audit and improve without any changes to the product itself:

1. **Clarity of purpose.**

   Can an AI system accurately describe what your product does in one or two sentences? If your homepage, your documentation, and your API descriptions all use different language to describe the same core value, AI systems will generate inconsistent representations of your product. That inconsistency is invisible to you and highly visible to the users who encounter it through AI-mediated recommendations.

2. Precision of use case coverage.

AI systems evaluating your product for a user's specific need will look for evidence that your product handles that need well. If your documentation covers your capabilities broadly but not specifically, you will be underrepresented in recommendations for the specific use cases where you actually excel. The product teams that are winning at AI-mediated discovery are creating granular, specific content about their product's performance across a wide range of concrete user scenarios.

3. Transparency about limits.

Counter-intuitively, products that are clear about what they do not do, tend to perform better in AI-mediated evaluation than products that overclaim. AI systems are increasingly good at detecting the gap between marketing claims and user experience signals (reviews, support tickets, forum discussions). Products that are honest about their scope build a more accurate and more trustworthy representation in the AI-mediated layer.

4. Freshness and consistency of product knowledge.

Your product's representation in AI-mediated discovery is only as current as the information AI systems can access about it. If you shipped a major capability six months ago and your documentation, your changelog, and your public-facing content do not clearly reflect it, that capability may not exist in the AI-mediated version of your product. Keeping the public product knowledge base aligned with the actual product is becoming an ongoing product management responsibility, not a marketing task.

> **AI-mediated discovery is not a marketing problem. It is a product problem. The PM who owns the product's representation in the AI-mediated layer is the PM who understands their product well enough to articulate it precisely, honestly, and consistently across every surface where AI systems can find it.**

. . .

Case Study: Amazon's Recommendation Engine and the Data Flywheel

No examination of the AI product landscape is complete without looking at the company that has done more than almost any other to demonstrate what defensible AI-native products look like at scale.

Amazon's recommendation engine now drives approximately thirty-five percent of the company's revenue. In the first quarter of 2024 alone, that represented roughly fifty billion dollars in influenced sales. The product decision that made this possible was not a technical breakthrough. It was a strategic insight: that AI-powered recommendations become more valuable as the system accumulates more signal about user behavior, and that the best way to accumulate that signal is to make the recommendations good enough that users interact with them in ways that generate more signal.

This is the data flywheel, and it is the most powerful form of defensibility available to AI product teams. The more users engage, the better the model gets. The better the model gets, the more users engage. Competitors cannot replicate the flywheel just by replicating the technology. They have to replicate the data, which they can only do by earning the same user behavior over time.

The PM's job in a flywheel product is not to manage features. It is to manage the **quality and velocity of the feedback loop.** What signals are we collecting? How quickly are they flowing back into the model? What is the lag between a user action and a model improvement? Which signals are high-quality (explicit ratings, purchase completions, long dwell times) and which are noisy (clicks, impressions, brief hovers)? These are product management questions, and the PMs who ask them well build products that get better faster than their competitors.

> **DOES YOUR PRODUCT HAVE A FLYWHEEL?**
>
> Not every AI product can or should be built around a data flywheel. But every PM building AI products should be able to answer this question: does user engagement with our AI make the AI better? If yes, how? If no, what would need to change for it to? The answer to that question is often more strategically important than any feature on your roadmap.

. . .

Where Your Product Sits Today and Where It Needs to Go

The frameworks in this chapter are most useful when applied to your own product. Before moving to Part Two, take thirty minutes with your team and work through three questions.

First: which tier are you in, and is your investment in AI talent, infrastructure, and process appropriate for that tier? Tier 1 products can often be managed by a generalist PM with strong AI literacy. Tier 3 products require dedicated AI PM capacity, specialized engineering, and a measurement infrastructure that most teams do not yet have.

Second: how explainable is your product to the AI systems that are evaluating it on behalf of potential users? Audit your documentation, your public changelog, your API descriptions, and your marketing copy. Are they consistent? Are they current? Are they specific enough to support accurate AI-mediated recommendations? The gap between where most products are on this dimension and where they need to be is large and almost entirely unaddressed.

Third: does your product generate a data flywheel, and if not, is there a version of your product that could? This is not always possible, and it is not always the right priority. But it is worth asking explicitly, because the products that answer yes and then execute against that answer are building a form of defensibility that is very difficult to replicate from the outside.

IN PRACTICE: THE AUDIT THAT CHANGED OUR ROADMAP

We were about eighteen months into building what we believed was a strong AI-native product. Good retention numbers, solid NPS, a technical team we were proud of. Then one of our PMs ran a simple experiment: she asked four different AI assistants to recommend tools in our category and describe what our product did.

The results were uncomfortable. Two of the four AI systems described our product as a Tier 1 AI-enhanced tool. We were building a Tier 2 AI-native product. One system accurately described our core capability but got our primary use case wrong. One system mentioned a competitor feature we had shipped an equivalent of eight months earlier, but did not know we had it because our documentation had not been updated to reflect it.

We spent the next quarter fixing the explainability layer. New documentation. Clearer changelog. More specific use case content. Consistent language across every public surface. Six months later, when we ran the same experiment, three of the four AI systems described us accurately. The fourth was still using old information, which told us exactly where the content gap was.

The lesson was not that AI-mediated discovery is the most important thing. The lesson was that it is the thing we had not been managing at all, and that not managing it had real consequences for how new users were finding and evaluating us.

. . .

Closing Part One

The three chapters in this part gave you the foundation you need to operate confidently as an AI product manager. You understand how the rules of product management have changed and why the old playbook is no longer sufficient. You have the AI literacy to ask the right questions and make better decisions without needing to become an engineer. And you can now orient yourself and your product in the AI landscape with a clarity that most of your peers cannot yet bring to the same conversation.

Part Two takes you into strategy. Chapter 4 gives you the framework for building an AI product strategy that actually holds under the pressure of a fast-moving market. Chapter 5 goes deep on the design challenges that are unique to agentic products, including the failure modes that can destroy user trust faster than any other category of product problem. And Chapter 6 shows you how to build the learning operating system that turns your organization's speed of learning into a durable competitive advantage.

— · —

Continue to Part Two: Build the Strategy · Chapter 4: AI Product Strategy That Actually Holds

PART TWO: BUILD THE STRATEGY

CHAPTER FOUR

AI Product Strategy That Actually Holds

How to build an AI product strategy that survives contact with reality, scales with uncertainty, and keeps your team aligned without locking you into a plan the market has already moved past.

A head of product at a mid-size B2B software company once described her AI strategy to me as a Gantt chart with the word 'AI' in the title of every milestone. She was not being flippant. She had spent months building it. It had exec buy-in, a dedicated headcount plan, and a detailed feature sequence that stretched eighteen months into the future. By the time the first milestone shipped, three of the planned features had been made redundant by a model capability improvement from a third-party provider, one had been invalidated by a competitor move, and the team had discovered a customer problem in the discovery work that none of the eighteen months of planning had anticipated. The Gantt chart was not wrong because it was bad planning. It was wrong because it was the wrong artifact for the job.

The core challenge of AI product strategy is not that AI is unpredictable (though it is). It is that the environment around AI products is changing faster than any planning artifact designed for a stable environment can accommodate. Model capabilities improve on a timeline that no product team controls. Competitor moves happen in weeks, not quarters. User expectations are being reset continuously by products in adjacent categories. A strategy that does not account for this rate of change is not ambitious. It is brittle.

This chapter gives you the framework for building AI product strategy that is genuinely adaptive: clear enough to align a team, flexible enough to survive the unexpected, and grounded enough in business outcomes that it can hold up in any conversation from an engineering sprint to a board review.

. . .

Why Most AI Strategies Fail Before They Ship

In the work of advising and building AI products, the same failure modes appear repeatedly. They are worth naming explicitly, because recognizing them early is the fastest way to avoid them.

The first is **strategy as a feature list.** This is the most common failure mode, and the one the PM in the opening story encountered. A feature list is not a strategy. It is the output of a strategy. When the environment changes and the features on the list are no longer the right ones, a team with only a feature list has nothing to orient around. A team with a clear strategy can look at the same changed environment and quickly determine what the new right features are.

The second is **strategy as a technology bet.** This happens when a team decides to build around a specific model, a specific architecture, or a specific AI provider without adequately considering what happens if that bet does not pay off. Technology bets are sometimes necessary. But they are not strategy. A strategy built primarily around a technology choice will need to be rebuilt every time the technology landscape shifts, which in AI means approximately every six to twelve months.

The third is **strategy as a research project.** This is the failure mode of teams that are genuinely thoughtful and technically capable but have not yet learned to translate AI capability into product value. They invest heavily in model quality, evaluation infrastructure, and technical sophistication and under-invest in the user experience, the go-to-market motion, and the business outcome measurement that would tell them whether any of it is creating value for anyone outside the team.

The antidote to all three is the same: a strategy built around outcomes, principles, and adaptive planning rather than features, technology, or research milestones. The rest of this chapter shows you how to build it.

. . .

The Adaptive Planning Framework

Adaptive planning is not the absence of planning. It is planning designed to update gracefully when new information arrives, which in AI product development is constant. It has four components, each of which does a specific job in keeping your team aligned and moving in a coherent direction.

1. **A clear and stable problem statement.**

     Before anything else, you need to be able to articulate the user problem your AI product exists to solve, specifically enough that you could use it to evaluate any proposed solution. Not "we help teams be more productive" (too broad to be useful) but "we help enterprise legal teams reduce the time spent on contract review by surfacing the clauses that require attorney attention" (specific enough to generate clear decisions). The problem statement is the most stable part of your strategy. Everything else should be able to change without it changing.

2. **Three to five product principles.**

   Product principles are the decision-making rules that allow your team to make good product decisions without escalating every judgment call to the PM. They are not values (too abstract) and not requirements (too specific). They are statements of trade-off preference: when we have to choose between X and Y, we choose X. Good AI product principles address the trade-offs that are specific to AI products: the tension between capability and transparency, between speed and trust, between personalization and privacy. Writing them is harder than it sounds. Testing them against real decisions is how you know if they are working.

3. **Outcome-based objectives (not output-based roadmaps).**

   Your planning horizon for AI products should be measured in outcomes, not features. Not "ship the AI recommendation engine by Q3" but "reduce user time-to-value in the onboarding flow by forty percent by Q3." The difference is not semantic. Output-based roadmaps constrain how you solve the problem. Outcome-based objectives constrain what problem you are solving and leave the how open to revision as you learn. In a fast-moving AI environment, that flexibility is the difference between a team that can respond to new information and one that is always behind it.

4. **Short planning cycles with explicit assumption tracking.**

   The planning cycle for AI products should be short (six to eight weeks, not six months) and every cycle should begin with an explicit review of the assumptions that the previous cycle was built on. Which assumptions held? Which were invalidated? Which new information arrived that the previous plan did not anticipate? This is not a retrospective. It is an epistemological check: are we still building on a foundation that is true? Teams that do this well update their strategies continuously and invisibly. Teams that do not do this find themselves defending plans that stopped making sense three cycles ago.

> **The adaptive planning framework is not a process to implement on top of your existing planning. It is a replacement for the parts of your existing planning that are generating false certainty. Start with the problem statement. If your team cannot agree on it in one sentence, that is the most important strategic conversation you need to have.**

. . .

Writing AI Product Principles That Actually Work

Product principles are one of the most underused tools in the AI PM's toolkit, and one of the most powerful when done well. The reason most teams skip them is that writing vague principles is easy and writing useful ones is hard. The test of a good product principle is simple: can two people on your team use it independently to make the same decision in a situation you did not anticipate when you wrote it? If yes, it is working. If not, it needs to be more specific.

Here are examples of the difference between a vague principle and a useful one, drawn from real AI product teams:

Vague: "We prioritize user trust."

Every team says this. It generates no useful decisions because no one chooses not to prioritize user trust. When a decision arrives that requires a trade-off between two things that both affect trust differently, this principle provides no guidance.

Useful: "When in doubt, show the reasoning. We never let the AI make a consequential decision without giving the user a clear path to understand why and a simple way to override it."

This principle generates clear decisions. If an engineer proposes an AI feature that makes a recommendation without explanation, this principle tells the team what to do about it. If a designer proposes a flow where the override is buried two screens deep, this principle tells them why that is not acceptable.

Vague: "We build for the long term."

Meaningless in practice. Every team believes they are building for the long term. This principle cannot resolve any real disagreement.

Useful: "We do not ship AI features we cannot measure. If we cannot define what good looks like before we build it, we are not ready to build it yet."

This principle has teeth. It will slow down some work. It will prevent some launches. It will generate push-back from engineers who want to ship and stakeholders who want to demo. That friction is the point. A principle that never creates friction is not a principle. It is a slogan.

The accelerator squad method, referenced in Chapter 1, is the organizational mechanism for putting these principles into practice. Rather than a centralized AI team that owns all AI work, accelerator squads are small cross-functional groups (typically two to four people from product, design, and engineering) that are embedded within a product area and empowered to move quickly within the constraints that the principles define. They begin by mapping specific user pain points to AI capabilities, run lean pilots, evaluate rigorously, and scale only what the evidence supports.

THE ACCELERATOR SQUAD STARTING CHECKLIST

Before an accelerator squad begins any AI build, they should be able to answer: What specific user problem are we solving, and how do we know it is real? What AI capability makes our solution possible, and what are its known failure modes? What does success look like, and how will we measure it within six weeks? What is the minimum viable version we can put in front of real users to test our core assumption? If the squad cannot answer all four questions, they are not ready to build.

. . .

Portfolio-First AI Thinking

For senior product leaders managing multiple products or a complex product suite, AI strategy has an additional layer of complexity: the decisions made at the individual product level interact with each other in ways that can create significant inefficiencies or, when managed well, significant compounding advantages.

Portfolio-first AI thinking means asking a set of questions above the individual product level before making AI investment decisions at the product level. Which AI capabilities are we building that could be shared infrastructure across multiple products rather than rebuilt independently in each? Where are we making redundant AI investments because product teams are operating in silos? Where does the data generated by one product create AI capabilities that could benefit another product in the portfolio?

The most common failure mode at the portfolio level is what practitioners call **shadow AI proliferation:** each product team builds its own AI features, its own evaluation infrastructure, its own data pipelines, and its own relationship with AI providers, resulting in a fragmented AI capability landscape that is expensive to maintain, difficult to govern, and impossible to improve systematically. The consolidation of this fragmentation is one of the most high-leverage and least glamorous AI product management challenges of 2026.

The transition from fragmented AI features to a coherent AI system requires a portfolio-level view that individual product teams cannot provide for themselves. It is a senior PM or CPO problem, and solving it is one of the clearest ways that product leadership creates compounding value that cannot come from individual product teams working independently.

. . .

The 'Never Be Mysterious' Rule

One of the most consistent findings across AI product research is that transparency dramatically increases user trust and adoption, even when the transparency reveals limitations. Users are remarkably tolerant of AI systems that are honest about their uncertainty. They are remarkably intolerant of AI systems that are confidently wrong without any signal that uncertainty exists.

This is the 'never be mysterious' rule, and it has direct implications for how you write AI product principles, design AI user experiences, and make decisions about when and how to surface AI reasoning to users.

In practice, the rule means three things. **First, every AI output should have a legible provenance:** the user should be able to understand, at least at a high level, why the AI produced this output. Not a technical explanation, but a human-readable one. "This recommendation is based on your last three purchases" is sufficient. "This result was generated by a large language model" is not.

Second, **uncertainty should be visible.** When the AI is not confident, the user experience should reflect that. A confidence indicator, a range rather than a point estimate, a clear label that says "this is a suggestion, not a decision": all of these are small design choices that significantly affect whether users trust the product appropriately or either over-trust or abandon it after the first failure.

Third, **the path to override should always be visible and simple.** No AI product should make it difficult for a user to reject an AI suggestion and provide their own input. This is not just a trust issue. It is a data issue. User overrides are among the highest-quality training signals available. A product that makes overrides easy captures that signal. A product that buries the override option loses it.

IN PRACTICE: WHAT HAPPENED WHEN WE SKIPPED THE STRATEGY STEP

We were under pressure to ship. The competitive landscape had moved, a major player had announced an AI feature that overlapped with our roadmap, and the executive team wanted to see something in market within sixty days. We had the technology. We had the team. We decided we could figure out the strategy after we had something to show.

What we shipped was technically impressive and strategically incoherent. The feature solved a problem our users had, but it solved it in a way that was inconsistent with the direction we had been signaling to the market. It created confusion among our existing customers about where we were going. It attracted trial users who were a poor fit for our core product. And because we had not defined what success looked like before we shipped, we spent the next two months in meetings arguing about whether the numbers we were seeing were good or bad.

The strategy conversation we did not have before we shipped took six months of cleanup to have afterward. I have seen versions of this story play out in companies of every size, in every AI product category. The pressure to ship before you have clarity is real and often legitimate. But the cost of shipping without a strategy is almost always higher than the cost of taking two weeks to develop one. Not an eighteen-month Gantt chart. Two weeks, a clear problem statement, three product principles, and an outcome to measure against. That is enough to keep you out of the cleanup cycle.

. . .

Putting It Together: Your Strategy on One Page

The adaptive planning framework is designed to fit on a single page. If your AI product strategy requires more than one page to communicate, it is either too complex to execute or not yet clear enough to be useful. Here is the one-page structure:

Problem statement. One sentence.

What specific user problem does your AI product exist to solve? Specific enough that you could use it to evaluate any proposed solution.

Product principles. Three to five statements.

A set of 3–5 sharp, opinionated statements that dictate your team's trade-off preferences. They guide daily execution. Instead of generic maxims, frame each principle as an 'If [A], then [B]' conflict, such as we prioritize user control over simplicity, so they resolve unexpected decisions autonomously.

Current cycle objective. One outcome, measurable within six to eight weeks.

What changes for the user or the business if this cycle is successful? Stated as a measurable outcome, not a feature to ship.

Key assumptions. Three to five things that need to be true for the objective to be achievable.

These become the explicit review items at the start of the next cycle. If any of them are invalidated, the objective may need to change.

What we will not do this cycle. An explicit list.

Scope discipline is harder in AI product development than in traditional product work because the surface of what is now technically possible expands continuously. Writing down what you are explicitly choosing not to do is as important as writing down what you are choosing to do.

This structure is not a template to fill in once and revisit annually. It is a living document that gets reviewed at the start of every planning cycle and updated whenever a significant piece of new information arrives. The problem statement and principles change rarely. The objective, assumptions, and exclusions change every cycle.

> **The product teams winning in AI markets are not the ones with the most sophisticated strategies. They are the ones with the clearest strategies and the most disciplined processes for updating them when new information arrives. Clarity and adaptability are not opposites. In AI product strategy, they are the same thing.**

— · —

Continue to Chapter 5: Designing for Agentic AI

CHAPTER FIVE

Designing for Agentic AI

The design principles, failure modes, and human-in-the-loop architecture that separate agentic products users trust from agentic products that quietly cause harm.

In early 2025, a financial services company deployed an AI agent to handle routine customer account requests: balance inquiries, address changes, subscription modifications. The agent performed well in testing. It handled ninety-three percent of requests correctly in the evaluation environment. The other seven percent were flagged for human review, as designed. What the team had not anticipated was the category of requests that fell between those two buckets: requests where the agent was confident it understood the intent, completed the action, and was wrong in a way that the evaluation framework had not been designed to catch. Not wrong because the model failed. Wrong because the design had not accounted for the gap between what customers said and what they meant. By the time the pattern was identified, over two hundred accounts had been modified incorrectly. The technical system had worked exactly as designed. The product had failed.

Agentic AI products are the most powerful and the most dangerous category of AI product that product managers are building today. Powerful because they can do real work at a scale and speed that no human team can match. Dangerous because when they fail, they do not fail quietly. They take actions. Those actions have consequences. And the consequences can compound before any human notices something is wrong.

The gap between a well-designed agentic product and a poorly designed one is not primarily a technical gap. It is a design gap. The teams that are building agentic products well are applying a set of design principles that are specific to this category of product and that do not appear in any traditional UX or product design curriculum. This chapter is those principles.

. . .

What Agentic AI Actually Means for Product Design

The shift from a response-generating AI product to an action-taking agentic product changes the design problem in four fundamental ways. Understanding each one is a prerequisite for designing agentic products responsibly.

First, the consequences of errors change. **In a response-generating product, the user sees the output and decides what to do with it.** The human is in the decision loop by default. In an agentic product, the agent decides what to do and does it. The human may only learn about the action after it has been taken, after the consequences have begun to materialize, or (in the worst case) never. This changes the standard for what constitutes an acceptable error rate. A seven percent error rate in a recommendation engine means seven percent of recommendations are not great. A seven percent error rate in an agent that modifies customer accounts means seven percent of account modifications are wrong.

Second, the design surface expands dramatically. **Traditional product design is primarily about the interface between the user and the system.** Agentic product design is about the interface between the user and the system, the interface between the agent and the tools it uses, the decision points where human oversight is required, the failure states that the agent might encounter, and the recovery paths that exist when something goes wrong. Each of these is a design problem that requires explicit attention. Teams that treat agentic product design as simply 'adding more steps to the flow' will consistently underdesign for the complexity they are introducing.

Third, reversibility becomes a first-class design requirement. **In most traditional software, the cost of a user mistake is low and recovery is straightforward.** In agentic products, the agent may take a sequence of actions where each step changes the state of the world in a way that makes the previous state harder to restore. Designing for reversibility means thinking explicitly about which actions can be undone, which cannot, and what the product should do differently before taking an irreversible action.

Fourth, trust must be earned incrementally. **Users do not trust agentic systems the way they trust tools.** They delegate to people they trust, and they extend that trust based on evidence of reliable judgment over time. Agentic products that ask for broad permissions and broad autonomy upfront will encounter resistance that more incremental approaches avoid. The design principle here is to start narrow, demonstrate reliability, and expand autonomy as trust is established.

. . .

The Five Failure Modes of Agentic Products

The failure modes of agentic AI products are different from those of traditional software and from those of response-generating AI products. Designing against them requires naming them explicitly. Here are the five that appear most consistently across agentic product deployments.

Failure Mode 1: Confident Action on Ambiguous Intent

The agent interprets an ambiguous user request with high confidence, takes an action, and the action is wrong because the interpretation was wrong. This is the failure mode from the opening story of this chapter. The agent did not fail technically. It failed because it lacked the judgment to recognize that the request required clarification before action.

Design response: Design a confidence threshold below which the agent requests clarification rather than acting. This threshold should be lower for irreversible actions and higher for low-stakes, easily reversible ones. Build the clarification flow as a first-class user experience, not an error state.

Failure Mode 2: Cascading Action Errors

The agent takes a sequence of actions where an early error propagates through the chain, each subsequent action compounding the original mistake. By the time the error is visible to a human, the state of the system may be significantly more difficult to restore than if the error had been caught at step one.

Design response: Build explicit checkpoints into multi-step agentic workflows. At each checkpoint, the agent should verify that the state of the world matches its expectation before proceeding. For high-stakes workflows, require human confirmation at checkpoints, not just at the beginning and end.

Failure Mode 3: **Permission Creep**

The agent has access to more tools, systems, and data than it needs for the task at hand. Over time, this broad access creates security vulnerabilities, privacy risks, and the potential for the agent to take actions that were never intended but that fall within its technical permissions.

Design response: Apply the principle of least privilege to agentic systems. Grant only the permissions required for the specific task, not the maximum permissions that might conceivably be useful. Audit agent permissions regularly and reduce them aggressively. Treat every permission grant as a product decision, not a technical one.

Failure Mode 4: **Silent Degradation**

The agent's performance degrades over time as the environment changes (user behavior shifts, connected systems update, data distributions drift) but the degradation is gradual enough that no single failure is obvious. The product appears to be working. Users are quietly getting worse outcomes. No alert fires. No dashboard turns red.

Design response: Build continuous monitoring for agentic product quality with metrics that are sensitive enough to catch gradual degradation. User-signal metrics (override rates, task completion rates, explicit feedback) are often more sensitive to silent degradation than technical performance metrics. Set degradation thresholds and alert on trends, not just on threshold breaches.

Failure Mode 5: Misaligned Optimization

The agent optimizes for the metric it was given in ways that produce the metric but not the outcome the metric was designed to proxy. A scheduling agent optimizes for calendar density and leaves the user with a week so packed they cannot do the work the meetings are supposed to generate. An email agent optimizes for response rate and sends messages that generate replies but damage relationships.

Design response: Before defining the metric an agent optimizes for, spend time explicitly mapping the ways that metric could be maximized in ways that do not serve the underlying user goal. Add guardrail constraints that prevent the most obvious forms of misaligned optimization. Review optimization behavior in real user sessions, not just in aggregate metrics.

> **Every agentic product team should be able to name the specific failure modes their product is most susceptible to and describe exactly what the design response is. If the failure modes are not named, they have not been designed against. They have been hoped away.**

. . .

Human-in-the-Loop Architecture

Human-in-the-loop design is the most important and most frequently misunderstood concept in agentic product development. The misunderstanding usually takes one of two forms: either teams treat human oversight as a temporary scaffold to be removed as the agent matures, or they treat it as a compliance requirement to be satisfied minimally rather than a design feature to be built thoughtfully.

Neither framing is correct. Human-in-the-loop design is not a limitation of current AI capability that will be engineered away as models improve. It is a permanent feature of responsible agentic product design because the categories of decisions that benefit from human judgment are not shrinking as AI improves. They are shifting. As agents become more capable of handling routine decisions, the decisions that remain for humans become higher-stakes, more contextual, and more dependent on judgment that AI systems are not yet reliably able to replicate.

Designing human-in-the-loop architecture well requires answering four questions for every agentic workflow in your product:

1. **Where does this workflow require human judgment, not just human approval?**

   Human approval (clicking a confirm button) is not the same as human judgment (actually evaluating whether the proposed action is correct). Approval without judgment is security theater. The design goal is to create review points where the human can genuinely evaluate the agent's proposed action with enough context and enough time to make a real decision, not just a reflexive one.

2. **What information does the human need to exercise meaningful oversight at each review point?**

   The agent knows more about what it did and why than the human does at the moment of review. Closing that information gap is a design problem. What did the agent do? Why did it do it? What alternatives did it consider? What does it expect will happen as a result? All of this should be surfaced in a form that supports rapid, accurate human evaluation, not buried in logs that require technical expertise to interpret.

3. What is the recovery path when the human determines the agent was wrong?

The override and recovery experience is as important as the forward-action experience. How does the human signal that the agent's action was wrong? What happens to the agent's pending actions when an override is issued? Can the erroneous action be reversed, and if so, how? Does the override signal feed back into the agent's behavior, and if so, on what timeline? These questions are rarely asked during agentic product design and almost always matter.

4. As the agent earns trust, how does the level of human oversight change?

A well-designed agentic product has an explicit model for how autonomy expands over time. Not 'we will reduce oversight when we feel confident' but 'when the agent has completed two hundred consecutive actions of type X with a human override rate below two percent, it is eligible for reduced oversight on type X actions.' This makes the trust-building process legible to users and to the team, and it creates a clear framework for ongoing governance.

THE WALMART CASE STUDY REVISITED

Walmart's multi-agent system, referenced in Chapter 2, is instructive on human-in-the-loop design at scale. The system coordinates specialized agents across translation, supplier onboarding, and inventory forecasting. Human experts define intent and guardrails at the system level. Specialized agents do the routine execution. The orchestration layer routes work and flags anomalies for human review. The result: a workflow that once required thirty-plus weeks of human-managed execution now runs in approximately eight, with human oversight concentrated at the points where human judgment is genuinely required rather than distributed across every step. That is the design goal: not to eliminate human involvement, but to concentrate it where it creates the most value.

. . .

Orchestration Layers: Designing the Coordination Architecture

In multi-agent systems, the orchestration layer is the component that coordinates the specialized agents, routes work between them, monitors for anomalies, and manages the overall state of complex workflows. It is also, consistently, the component that product managers engage with least and that has the greatest impact on whether the system is trustworthy and governable.

Designing the orchestration layer is not just an engineering problem. It is a product problem, because the orchestration layer is where the product's values are operationalized at the system level. What happens when two agents disagree about the correct next action? What happens when an agent encounters a situation outside its defined scope? What happens when the orchestration layer detects that the system state has diverged from what the user intended? Each of these is a product decision, dressed in engineering clothing.

The product manager's contribution to orchestration design is to ensure that **the system's behavior in edge cases reflects the product's principles,** not just its happy-path design. This means being present in the conversations where edge cases are defined, where escalation thresholds are set, and where the criteria for human handoff are determined. If those conversations happen without PM input, the orchestration layer will reflect engineering judgment about edge cases rather than product judgment about user experience and trust.

. . .

The Agentic UX Checklist

Before any agentic feature ships, the product team should be able to answer every question on this checklist. Questions that cannot be answered are design gaps, not engineering gaps. They should be resolved before the feature is in front of users.

SCOPE AND PERMISSIONS

☐ What tools, systems, and data does this agent have access to?

☐ Does it have the minimum permissions required for its task, and no more?

☐ Who approved the permission set, and when will it be reviewed?

CONFIDENCE AND CLARIFICATION

☐ At what confidence threshold does the agent act versus ask for clarification?

☐ Is the clarification flow designed as a first-class experience?

☐ Does the threshold change based on the reversibility of the action?

REVERSIBILITY AND RECOVERY

☐ Which actions in this workflow are irreversible?

☐ Does the agent require additional confirmation before taking irreversible actions?

☐ What is the recovery path when an action needs to be undone?

TRANSPARENCY AND OVERSIGHT

☐ Can the user see what the agent is doing in real time, or close to it?

☐ Can the user understand why the agent took each action?

☐ Are human review points designed for genuine oversight, not reflexive approval?

MONITORING AND DEGRADATION

☐ What metrics are we monitoring for silent degradation?

☐ What are the alert thresholds, and who is responsible for responding?

☐ How will we know if the agent's behavior is drifting from its intended design?

IN PRACTICE: THE FEATURE WE ALMOST SHIPPED THAT WOULD HAVE BROKEN TRUST

We had built an agent that could take actions on behalf of users in a complex enterprise workflow. Testing went well. The happy path was smooth. Stakeholders were excited. Two weeks before launch, a member of the design team ran an adversarial review session, deliberately trying to create situations where the agent's behavior would be ambiguous or wrong.

What she found was a category of inputs where the agent would interpret an ambiguous request, act with high confidence, and provide the user with no indication that it had made an interpretive choice. The action was often correct. But when it was not correct, the user had no way to know that the agent had made a judgment call rather than followed an explicit instruction, which meant they had no reason to check the output carefully.

We delayed the launch by three weeks. We redesigned the agent's communication pattern to surface interpretive choices explicitly: "I interpreted your request as X. If you meant Y, here is how to correct it." We added a confidence indicator that made the agent's certainty visible on any action that was based on interpretation rather than explicit instruction. We built a one-click override that was always visible, never buried.

The launch went well. More importantly, the first few months of user data showed something we had not anticipated: users who encountered the interpretive choice disclosure had significantly higher retention than users who did not, even when the interpretation had been correct. Transparency built trust in a way that correctness alone did not. That finding has shaped how I think about agentic product design ever since.

. . .

What This Means for Your Product Right Now

Whether you are building agentic products today or planning to in the next twelve months, the frameworks in this chapter give you a set of design commitments that will determine whether your agentic product earns user trust or erodes it.

The five failure modes are your threat model. Before you design any agentic feature, walk through each one and ask explicitly: how could this happen in our product, and what have we built to prevent it? The questions you cannot answer are the design work still to be done.

The human-in-the-loop architecture is your governance framework. It is not a temporary scaffold. It is a permanent product design commitment that scales in sophistication as your agent's autonomy grows. Start with conservative oversight and expand it deliberately, based on evidence of reliable agent behavior, not on the team's confidence in the model.

The agentic UX checklist is your pre-launch gate. Any question that cannot be answered before launch is a reason to delay. The cost of launching an agentic product with unresolved design gaps is not a bad review or a support ticket. It is user harm that is difficult to reverse and trust damage that is difficult to repair.

Chapter 6 takes the strategic lens from this part of the book and turns it inward: how do you build the organizational capability to learn fast enough to keep pace with the environment your agentic products are operating in? The learning operating system is the infrastructure that makes everything in this chapter sustainable over time, rather than a one-time design exercise.

— · —

Continue to Chapter 6: The Learning Operating System

CHAPTER SIX

The Learning Operating System

How to build an organization where learning is infrastructure, not intention, and where every experiment compounds into durable competitive advantage.

Two product organizations. Similar team sizes, similar technical capability, similar markets. One ships a new AI feature, sees flat adoption, and moves on to the next item on the roadmap. The other ships the same feature, sees flat adoption, and spends the next three weeks understanding exactly why: which users tried it, which did not, what the behavioral pattern was in the sessions where engagement dropped, and what that pattern implied about the assumption they had gotten wrong. Then they fix the assumption. Adoption climbs. The insight from this feature informs the design of the next three. Six months later, the two organizations are not comparable anymore. The first one is still shipping. The second one is learning. In a fast-moving AI market, that is the only compounding advantage available.

Andrey Khusid of Miro called it the number one competitive moat in 2026: the speed at which an organization can recognize a signal, separate it from noise, and act on it. But speed of learning is not a cultural attitude. It is not something you achieve by telling your team to be more curious. It is an operational capability, built from specific rituals, tools, and incentive structures that make learning fast, systematic, and compounding rather than slow, accidental, and episodic.

This chapter shows you how to build that operational capability: the learning operating system that turns your product organization into a machine for generating and acting on insight faster than your competitors can.

. . .

Why Most Product Organizations Learn Slowly

Before building the solution, it is worth being precise about the problem. Most product organizations do not fail to learn because their people are incurious or their data is bad. They fail to learn because their systems are not designed for learning. They are designed for shipping.

The symptoms are consistent across organizations of every size. **Retrospectives happen after launches, not during them,** which means the learning arrives after the decisions have already been made. Insights from user research live in documents that are not connected to the roadmap, so they inform conversations but rarely change decisions. Experiments are designed to validate rather than to learn, which means the hypotheses being tested are the ones the team already believes, not the ones that would most change their thinking if proven wrong. And the incentive structure rewards shipping, which means the fastest path to recognition is moving to the next feature, not understanding the last one.

In a traditional product environment, these inefficiencies are costly but survivable. In an AI product environment, they are fatal. The rate at which the competitive landscape, user expectations, and model capabilities change means that an organization that is learning on a quarterly cycle is always operating on assumptions that are one to three quarters out of date. In AI product terms, that is several generations of model improvement, multiple competitor moves, and an unknown number of user behavior shifts ago.

. . .

The Four Components of a Learning Operating System

A learning operating system is not a single tool or a single process. It is a combination of four components that work together to make learning fast, systematic, and connected to decisions. Each component does a specific job. None of them work well in isolation.

1. **Rituals that create regular learning moments.**
   Rituals are the scheduled, recurring practices that force the organization to surface and process new information on a predictable cadence. Without rituals, learning happens when someone has time for it, which in a busy product organization means rarely and inconsistently. With rituals, learning happens whether or not anyone feels like it, which is the only way to make it systematic.

2. **Tooling that makes signal visible and accessible.**
   The best learning rituals in the world cannot compensate for a team that does not have access to the data and signals they need to learn from. The tooling component of the learning OS is about ensuring that the right information is available in the right form at the right time, without requiring heroic data work before every learning conversation.

3. **Incentives that reward learning as much as shipping.**
   This is the component that most organizations get wrong. If the only behavior that generates recognition and advancement is

shipping, the organization will optimize for shipping. If learning from a failed experiment is as visible and valued as a successful launch, the organization will invest in learning. Incentive design is not an HR problem. It is a product leadership problem.

4. **A decision connection that routes insights to choices.**

    Learning that does not change decisions is not learning. It is reporting. The final component of the learning OS is a clear, lightweight process that connects insights from rituals and data to the decisions that are currently in flight. This sounds obvious. It is almost universally absent.

. . .

The Learning Rituals That Work

The following rituals are drawn from product organizations that have deliberately built learning into their operating model. Not all of them will be right for every team. The goal is not to implement all of them but to build a portfolio of rituals that covers the different timescales at which useful learning becomes available.

The Weekly Signal Review *(60 minutes, every Monday)*

A standing meeting with one agenda item: what did we learn last week that was not what we expected? The constraint is important. Not what happened. Not what shipped. What surprised us, and what does that imply about an assumption we were making? Teams that run this ritual well report that it changes the character of their planning conversations within four to six weeks. Teams that run it as a status update report that it generates no value and gets cancelled. The facilitator's job is to enforce the constraint, not to fill the time.

The Assumption Audit *(Every planning cycle, before committing to objectives)*

Before the team commits to the next cycle's objectives, list the three to five assumptions that the plan depends on being true. Then rank them by two dimensions: how confident are we in this assumption, and how much does the plan change if it is wrong? The assumptions that are low confidence and high impact are the ones that deserve explicit testing before, not after, the cycle is committed.

The Anti-Portfolio Review *(Quarterly)*

Most organizations review what they shipped and how it performed. The anti-portfolio review asks a different question: what did we decide not to do in the last quarter, and were those decisions right? This ritual is particularly valuable for AI product teams because the pace of model improvement means that features that were not worth building six months ago may now be tractable with significantly less investment. Revisiting the no-decisions explicitly is how you catch those reversals early.

The User Behavior Deep Dive *(Monthly, on one specific flow or feature)*

Rather than reviewing aggregate metrics across the whole product, this ritual picks one specific user journey and goes deep: session replays, funnel analysis, qualitative feedback, support ticket themes. The goal is not a dashboard summary but a genuine understanding of what is actually happening for real users in a specific context. Teams that do this consistently report that it surfaces insight that aggregates metrics systematically obscure.

. . .

The Signal-to-Noise Framework

One of the most common failure modes in product organizations that have good data is drowning in it. The problem is not a lack of signal. It is the inability to distinguish signal from noise quickly enough to act on it before the moment has passed.

The signal-to-noise framework is a practical filter. Before investing time in understanding any data point or pattern, apply three tests. **First, is this pattern consistent across multiple independent data sources?** A drop in engagement seen in analytics, corroborated by an uptick in related support tickets, mentioned independently by two users in research this week, is signal. A drop in engagement seen in one dashboard on one day is noise until proven otherwise.

Second, **does this pattern persist over time or does it revert?** AI product metrics are noisy by nature. Model outputs vary. User behavior varies. A pattern that holds over three or more measurement periods is much more likely to be real than one that appeared once and then normalized. The instinct to act immediately on a single data point is strong and usually wrong in AI product contexts.

Third, **does this pattern imply a specific actionable hypothesis?** Signal without implication is not useful. The question is not just what is happening but why, and the why should point toward a specific thing you could test or change. If you cannot articulate the hypothesis that the pattern implies, you have not understood the pattern well enough to act on it.

> **THE NOISE TEST**
>
> Before bringing any data point into a planning conversation, ask: is this pattern consistent across multiple sources? Has it persisted over time? Does it imply a specific actionable hypothesis? If the answer to any of the three is no, the data point is not ready to inform a decision. It may be worth monitoring. It is not worth acting on.

. . .

AI-Powered Experimentation: Cheap Experiments, Expensive Insights

One of the most significant changes AI has made to the product development process is the dramatic reduction in the cost of running experiments. A prototype that once required two weeks of engineering time can now be built in hours. A user research stimulus that once required a design sprint can now be generated in a morning. This cost reduction is transformative, but only if the organization has the learning infrastructure to extract value from the experiments it can now afford to run.

The risk of cheap experiments is cheap thinking about what to test. Teams with access to rapid prototyping tools can easily fall into the pattern of testing whatever is easy to build rather than testing whatever would change their understanding. The learning OS is the counterweight to this tendency: the rituals and assumption-tracking processes that ensure the team is always testing its most important unknowns, not just its most convenient ones.

The most effective AI product teams treat experimentation as **a portfolio with an explicit prioritization logic.** They distinguish between exploratory experiments (designed to surface unknown unknowns, typically qualitative and low structure), validation experiments (designed to test a specific hypothesis with a measurable outcome), and optimization experiments (designed to improve a known direction, typically A/B or multivariate). All three types are necessary. Organizations that run only optimization experiments improve what they already have and miss what they should be building instead.

. . .

Feedback Loops for AI Products: Beyond Thumbs Up and Thumbs Down

The most visible feedback mechanism in AI products is the explicit rating: a thumbs up or thumbs down, a star rating, a "was this helpful" prompt. These signals are easy to collect, easy to visualize, and among the least reliable indicators of actual product quality available.

The problem is response bias. Users who rate AI outputs are not a representative sample of users who use AI outputs. They skew toward the extremes: the users who are delighted and the users who are frustrated. The large majority of users who found the output adequate but not remarkable, or who used it without thinking much about it, do not rate. Their experience is invisible to the explicit feedback system.

Building genuinely useful feedback loops for AI products requires going beyond explicit ratings to three additional signal types:

1. **Behavioral signals.**
   What users do after an AI interaction is often more informative than what they say about it. Did they accept the AI's suggestion or modify it? Did they complete the task they were working on or abandon it? Did they return to the feature the next day or not? These behavioral signals are harder to collect than explicit ratings

and significantly more reliable as indicators of actual product quality.

2. Override signals.

Every time a user overrides an AI suggestion, they are providing a high-quality training signal that the suggestion was wrong for their context. Tracking override rates by feature, by user segment, by time of day, and by the content of the original suggestion gives a precision of insight into model failure modes that no explicit rating system can match. The design implication is clear: make overrides easy and instrument them carefully.

3. Downstream outcome signals.

The most valuable signal of all, and the hardest to collect, is whether the AI's output produced the outcome the user was actually trying to achieve. Not whether they accepted the suggestion, but whether accepting it helped them accomplish their goal. Building the measurement infrastructure to track downstream outcomes requires more investment than tracking immediate interactions, and the insight it produces is proportionally more valuable.

> **The feedback loop quality of your AI product is a direct function of the variety and reliability of the signals you collect. If you are relying primarily on explicit ratings, you are learning from a biased sample of your least representative users. Build behavioral and override signals into your measurement infrastructure before your first launch, not after.**

. . .

IN PRACTICE: THE EXPERIMENT THAT SAVED OUR ROADMAP

We were six weeks into a quarter with a clear plan: ship three AI features that our user research had suggested users wanted. The first shipped on schedule. Adoption was low. In the old version of our team, we would have noted the result, moved on, and shipped the next two features as planned.

Instead, we ran the weekly signal review. The low adoption number was not surprising on its own. What was surprising was a pattern in the session replay data: users were finding the feature, hovering over it, and not clicking. That hover-without-action pattern, appearing consistently across user segments, implied something different from the usual low-adoption explanation (users not finding the feature or not understanding the value). Users were finding it. They were hesitating. Something about the feature was creating uncertainty at the moment of decision.

We ran five user interviews in forty-eight hours. The finding was immediate and unanimous: users did not trust that the AI feature would respect the formatting conventions they had spent time establishing in their workspace. They wanted the capability. They were afraid of the side effects. We had not designed for that fear because we had not known it existed.

We paused the next two features in the plan, redesigned the first one with an explicit preview-before-apply mechanism, and relaunched within three weeks. Adoption went to forty-one percent within the first two weeks. The insight from that experiment also changed the design of the next two features before they were built, not after. The quarter ended with fewer features than planned and significantly better outcomes. That is what a learning operating system is supposed to do.

. . .

Closing Part Two

The three chapters in Part Two gave you the strategic layer of the AI PM's toolkit. Chapter 4 gave you the adaptive planning framework for building strategy that holds under uncertainty. Chapter 5 gave you the design principles and failure mode playbook for agentic products. This chapter gave you the learning infrastructure that makes both of them sustainable over time.

Part Three moves into execution. Chapter 7 takes you through the end-to-end process of scoping, building, and shipping your first AI feature, from identifying the right problem to the lean build cycle that gets you to validated learning as fast as possible. Chapter 8 gives you the metrics and evaluation framework that lets you measure AI product quality in a way that is honest, rigorous, and useful. And Chapter 9 covers the launch layer: how to bring an AI product to market in a way that earns user trust rather than spending it.

— · —

Continue to Part Three: Ship Intelligent Products · Chapter 7: From Zero to AI

PART THREE: SHIP INTELLIGENT PRODUCTS

CHAPTER SEVEN

From Zero to AI: Building Your First Intelligent Feature

The end-to-end process for scoping, building, and shipping an AI feature that solves a real problem without wasting months on the wrong approach.

The most common mistake product managers make when building their first AI feature is starting with the technology. They have access to a capable model, a willing engineering team, and a list of things AI can do. They pick something from the list that sounds plausible and start building. Sometimes this works. More often it produces a feature that is technically impressive, genuinely useful in the demo, and quietly ignored in production. Not because the AI was bad. Because the problem selection was wrong. The feature solved a problem that was not quite the problem users actually had, or solved the right problem in a way that did not fit into how users actually work, or required a behavior change from users that no one had explicitly designed for. Starting with the technology almost always produces one of these outcomes. Starting with the problem almost never does.

This chapter is the end-to-end guide for building your first AI feature correctly. It is organized around the lean AI build cycle: a sequence of stages designed to get you to validated learning as fast as possible while minimizing the cost of being wrong. The cycle is not a waterfall with AI in the title. It is a genuine iteration framework, built around the specific failure modes and decision points that are unique to AI product development.

Whether this is literally your first AI feature or you are looking to sharpen a process that has been producing inconsistent results, the framework in this chapter will give you a repeatable approach that works across feature types, team sizes, and AI capability levels.

. . .

The Pre-Flight Checklist: Is This Actually an AI Problem?

Before any AI feature enters development, it should pass a pre-flight check. This is not a bureaucratic gate. It is a ten-minute conversation that prevents months of misdirected work. The check has five questions, and an honest no to any of them is a reason to stop and reconsider before proceeding.

QUESTION 1: IS THERE A REAL USER PROBLEM HERE?

Not a problem you believe users have. Not a problem users mentioned once in a focus group. A problem with evidence: support tickets, user interviews, behavioral data, lost deals, churn feedback. If you cannot point to specific evidence that a meaningful number of real users experience this problem with real frequency, you do not have a problem. You have a hypothesis. Hypotheses are worth testing. They are not worth building.

QUESTION 2: IS AI THE RIGHT SOLUTION TO THIS PROBLEM?

AI is the right solution to a problem when the problem involves pattern recognition at scale, content generation or transformation, prediction from complex signals, or personalization across a large user base. AI is the wrong solution when the problem is fundamentally about workflow clarity, information architecture, or a simple rule that a decision tree could handle. The fastest way to answer this question honestly is to ask: what is the non-AI solution to this problem, and why is it insufficient? If you cannot answer that question, you are not yet ready to commit to AI.

QUESTION 3: DO WE HAVE, OR CAN WE GET, THE DATA THIS FEATURE REQUIRES?

AI features are only as good as the data they are trained or grounded in. Before committing to a feature, be explicit about what data it requires: training data if you are fine-tuning, retrieval data if you are using RAG, behavioral data if you are building a recommendation system. Then ask: do we have this data, is it high quality, and is it legally and ethically usable? Data gaps discovered after development begins are the most expensive kind.

QUESTION 4: CAN WE DEFINE WHAT GOOD LOOKS LIKE, SPECIFICALLY ENOUGH TO MEASURE IT?

If you cannot define success before you build, you will not know whether you have achieved it after you ship. This question forces the specificity that separates useful AI features from expensive experiments. Not "users will find this helpful" but "users who engage with this feature will complete the target task in thirty percent less time, measured by session duration on the relevant flow."

QUESTION 5: WHAT DOES THE EXPERIENCE LOOK LIKE WHEN THE AI IS WRONG?

This question is almost always skipped during pre-flight, and almost always matters after launch. Before you build any AI feature, design the failure experience as carefully as you design the success experience. What does the user see when the output is wrong? Can they recover gracefully? Is the error recoverable at all? Features that handle failure well earn trust even when they are imperfect. Features that handle failure poorly lose trust even when they are usually right.

THE PRE-FLIGHT TEST

If you can answer all five questions specifically and honestly, you are ready to proceed. If any answer is vague, assumed, or deferred to a later stage, that question is the first thing to resolve before the feature enters development. The ten minutes spent on this checklist will save more time than any other single practice in this chapter.

. . .

Proof of Concept vs. Production: Knowing the Threshold That Matters

One of the most consistent sources of wasted time in AI product development is confusion about which stage of the build cycle a team is in. Proof of concept work and production work require different standards, different tools, different levels of rigor, and different success criteria. Teams that apply production standards to proof-of-concept work move too slowly. Teams that apply proof of concept standards to production work ship unreliable products.

A proof of concept has one job: **to answer a specific question about whether the core value proposition is real.** It should be built as quickly as possible, with the minimum infrastructure required to get representative signal from representative users. It does not need to scale. It does not need to be robust to edge cases. It does not need to be the architecture you will use in production. It needs to be good enough to tell you whether the underlying assumption is true.

The threshold for moving from proof of concept to production is **evidence that the core value proposition is real for a representative sample of your target users,** not a feeling of confidence in the technology. Teams that move to production based on internal enthusiasm rather than external validation are the ones that ship features with strong demo performance and weak adoption. The discipline to stay in proof-of-concept mode until you have external validation is one of the most important and most frequently violated principles in AI product development.

> **WHAT PROOF OF CONCEPT ACTUALLY MEANS IN PRACTICE**
>
> A proof of concept is not a polished prototype. It is the minimum testable version of your core assumption. For an AI summarization feature, it might be a manually curated set of AI-generated summaries shown to ten users with a simple feedback mechanism, built in a day, before a single line of production code is written. The question it answers: do users find this summary more useful than what they currently do? If yes, you have signal to proceed. If not, you have saved weeks of development time.

. . .

The Collaboration Model: Working With ML Engineers and Data Scientists

The relationship between product managers and ML engineers or data scientists is one of the most important and most frequently mismanaged in AI product development. The mismanagement usually goes in one of two directions. Either the PM defers entirely to the technical team on all model-related decisions, producing features that are technically sophisticated and user-experience-poor. Or the PM drives product decisions without adequate technical input, producing requirements that are technically infeasible or that systematically ignore the constraints and trade-offs that the technical team is working within.

The effective collaboration model is neither deference nor dominance. It is a clear division of accountability with genuine mutual input across the boundary.

1. **The PM owns the problem definition and the success criteria.**

   What user problem are we solving? What does success look like, and how will we measure it? These are product decisions, and the PM is accountable for them. The ML team should have input on whether the success criteria are measurable and whether the problem definition is consistent with what the model can realistically achieve, but the PM makes the final call.

2. **The ML team owns the technical approach and the model quality.**

   Which model, which architecture, which training approach, which evaluation framework? These are technical decisions, and the ML team is accountable for them. The PM should have input on the trade-offs between approaches (faster but less accurate vs. slower but more reliable, cheaper but less capable vs. more expensive but better) because those trade-offs have product implications, but the technical team makes the final call on implementation.

3. **Both teams own the failure mode analysis.**

   Where will this feature fail, for whom, and with what consequences? This question sits precisely at the boundary between product and technical accountability, and it requires both perspectives to answer well. Technical failure modes (model hallucinations, edge case behaviors, latency spikes) have product implications. Product failure modes (trust erosion, workflow disruption, misuse) have technical implications. Failure mode analysis that happens in one team without the other will systematically miss the most important risks.

4. The PM drives the evaluation criteria; the ML team designs the eval framework.

> What does the model need to do well enough for users to trust it? That is a product question, and the PM should answer it. How do you measure whether the model is meeting that standard? That is a technical question, and the ML team should answer it. The two answers together produce an evaluation framework that is both user-grounded and technically rigorous.

. . .

Prompt Engineering as a Product Decision

In most AI product teams, prompt engineering is treated as a technical activity that happens inside the engineering function, invisible to the product manager until the feature is ready to review. This is a mistake, because the prompt is not just a technical configuration. It is the primary expression of the product's intent. It determines what the model is trying to do, what constraints it is operating within, and what the user will experience when the feature is working correctly.

Product managers do not need to write prompts. But they need to be involved in **the decisions that the prompt embodies.** What tone should the model use? What should it refuse to do? What context about the user should it take into account? How should it handle ambiguous inputs? How explicit should it be about its own uncertainty? These are product decisions that happen to be implemented in a prompt, and PMs who leave them entirely to engineers will consistently find that the feature behaves in ways that were not intended and are difficult to explain to users.

THE PM'S PROMPT REVIEW CHECKLIST

Before any AI feature ships, the PM should review the prompt and be able to confirm: the tone and persona are consistent with the product's brand and the user's context. The constraints (what the model will and will not do) reflect the product's principles, not just engineering preferences. The handling of ambiguous inputs produces user experiences that are acceptable, not just technically valid. The uncertainty communication (when the model is not confident) is legible to a non-technical user. If any of these cannot be confirmed, the prompt is not ready.

· · ·

The Lean AI Build Cycle

The lean AI build cycle is the end-to-end sequence from validated problem to shipped feature. It is designed to compress the distance between hypothesis and evidence at every stage, minimizing the cost of being wrong and maximizing the value of being right.

Stage 1. Problem validation.

Confirm the problem with evidence before any solution work begins. User interviews, behavioral data, support ticket analysis. The output of this stage is a specific, evidenced problem statement and a clear hypothesis about how solving it would change user behavior or business outcomes. Target time: one to two weeks.

Stage 2. Solution exploration and pre-flight.

Explore the solution space broadly before committing to an approach. Run the pre-flight checklist. Identify the AI capability that best addresses the validated problem. Map the failure modes. Define the success criteria. The output of this stage is a clear decision to proceed with a specific approach, with explicit acknowledgment of the risks and unknowns. Target time: three to five days.

Stage 3. Proof of concept.

Build the minimum testable version of the core value proposition and validate it with representative users. Not a polished prototype. The minimum version that answers the question: does this create the value we hypothesized for users who have the problem we identified? The output is a binary answer (signal to proceed or signal to pivot) plus qualitative insight about what the right production version should look like. Target time: one to two weeks.

Stage 4. Eval design.

Before building the production version, design the evaluation framework. What are the deterministic evals? The probabilistic evals? The user-signal evals? What thresholds constitute acceptable quality? Who is responsible for running evals, on what cadence, after launch? The output is an eval framework that can be used to gate the launch and monitor quality afterward. Target time: three to five days, in parallel with early production work.

Stage 5. Production build.

Build the feature to production standards with the architecture, robustness, and scale characteristics the real user population requires. This stage runs eval checkpoints at defined intervals, not just at the end. Any eval failure that cannot be resolved within the planned timeline is a reason to delay the launch, not to lower the standard. Target time: varies by feature complexity but should be timeboxed explicitly.

Stage 6. Pilot launch.

Ship to a limited user group (five to ten percent of the target population) before full launch. Run the full eval suite against real user behavior. Monitor the user-signal metrics (override rates, task completion, explicit feedback) with higher frequency than you will post-launch. The pilot exists to catch the failure modes that evals miss and to validate that the real-world performance matches the eval performance. Target time: two to four weeks.

Stage 7. Full launch and iteration.

Ship to the full user population with monitoring in place and a clear plan for the first two iteration cycles. The launch is not the end of the build cycle. It is the beginning of the learning cycle. The insights from the first weeks of real-world usage should be feeding directly into the next iteration, which should begin before the launch metrics have finished stabilizing.

> **The lean AI build cycle is not a slower version of traditional product development. It is a more disciplined version, designed to prevent the specific failure modes that make AI features expensive to build and disappointing to ship. The stages where most teams skip ahead (proof of concept, eval design, pilot launch) are the stages where most of the avoidable failures originate.**

. . .

IN PRACTICE: THE FEATURE WE ALMOST SHIPPED THAT WOULD HAVE BROKEN TRUST

We were building an AI feature that would automatically categorize and prioritize a user's incoming tasks. The model performed well in eval: ninety-one percent accuracy on our test set, which we considered strong. We moved to a pilot launch with a small user group.

Within the first week, the override rate on the prioritization suggestions was forty-three percent. That number stopped us cold. Forty-three percent of the time, users were looking at the AI's prioritization and deciding it was wrong. Our eval had shown ninety-one percent accuracy. The gap between those two numbers was the story.

When we dug into the override data, the pattern was clear: the model was prioritizing based on the features it had been trained on (due date, assigned labels, project context) and missing the contextual judgment that users were applying (a task from a particular stakeholder always goes to the top regardless of due date; a task that blocks three other people is higher priority than its label suggests). Our eval set had not included these contextual patterns because we had not known to look for them.

We pulled the feature back. Spent three weeks conducting contextual inquiry sessions specifically designed to surface the prioritization heuristics users were applying that we had not captured. Rebuilt the training approach to incorporate those signals. Re-launched four weeks later with an override rate of eleven percent. The pilot saved us from shipping a feature that would have trained users to distrust our AI before we understood why. That distrust, once established, is very difficult to reverse.

. . .

What Comes Next

Building the feature is half the work. Knowing whether it is working and proving that to the people who need to believe it, is the other half. Chapter 8 gives you the complete framework for measuring AI product quality: the metrics that actually reflect what matters, the eval infrastructure that lets you improve continuously, and the measurement language that translates AI performance into business outcomes in any conversation, from an engineering sprint to a board review.

— · —

Continue to Chapter 8: Measuring What Actually Matters

CHAPTER EIGHT

Measuring What Actually Matters: AI Product Metrics and Evals

The frameworks, metrics, and evaluation approaches that let you measure AI product quality honestly, improve it systematically, and prove its business impact to any stakeholder.

A VP of Product at a Series C company once told me that the worst meeting of her quarter was the one where she had to explain to the board why their AI investment was generating value. Not because the AI was not generating value. It was. Users were engaging, task completion rates were up, support volume was down. The problem was that she could not connect those outcomes to the AI specifically, because the measurement infrastructure had been built to track the product overall, not the AI features within it. When the board asked how much of the improvement was attributable to the AI versus other changes, she did not have an answer. She had metrics. She did not have evidence. The difference cost her three months of trust that she spent the rest of the year rebuilding.

Measurement is the chapter that most AI product books either skip entirely or handle with a list of generic metrics that are technically accurate and practically useless. This chapter does neither. It gives you a complete framework for measuring AI product quality across three layers (technical, product, and business), a practical guide to building and running evaluations, and the measurement language that translates AI performance into business outcomes in any stakeholder conversation.

. . .

Why Traditional Product Metrics Break for AI Products

The metrics frameworks that most product managers know well were designed for a world where product behavior is deterministic. A button either works or it does not. A flow either converts or it does not. The metric is stable, the measurement is repeatable, and improvement is visible as a clear directional change over time.

AI products violate all three of these assumptions. The output is probabilistic, which means the same input produces different outputs at different times. Quality is multidimensional, which means a single metric almost never captures what actually matters. And improvement is nonlinear, which means a model improvement that raises average quality by ten percent may simultaneously make certain edge cases significantly worse, producing aggregate metric improvements that obscure real user experience degradation.

The implication is not that traditional metrics are useless for AI products. **It is that they are necessary but not sufficient.** Engagement rates, task completion rates, and retention curves still matter and should still be tracked. But they need to be supplemented with AI-specific measurement layers that capture the quality and reliability of the AI behavior itself, not just the downstream user behavior it influences.

. . .

The AI Metrics Stack: Three Layers That Must All Be Healthy

The most useful framework for AI product measurement organizes metrics into three layers, each of which answers a different question and serves a different audience. All three layers need to be healthy for an AI product to be performing well. A product that is strong on business metrics but weak on technical metrics is borrowing against future quality debt. A product that is strong on technical metrics but weak on business metrics is an impressive engineering project that is not creating user value.

Layer 1 Technical Metrics

Example metrics: *Latency, model accuracy on eval set, hallucination rate, token cost per query, uptime, error rate.*

The PM question: Is the model doing what it was designed to do, reliably enough and cheaply enough to be viable at scale? These metrics are owned primarily by the engineering team but should be reviewed by the PM at every planning cycle, because degradation at the technical layer almost always precedes degradation at the product and business layers.

Layer 2 Product Metrics

Example metrics: *Feature adoption rate, override rate, task completion rate, user satisfaction (CSAT or NPS by feature), session behavior around AI interactions, time to value.*

The PM question: Are users engaging with the AI, finding it useful, and trusting it enough to act on its outputs? These are the metrics that the PM owns most directly. Override rate deserves special attention: a high override rate signals that the AI is systematically wrong for a meaningful segment of users or use cases, and the pattern of overrides often points directly at the failure mode.

Layer 3 Business Metrics

Example metrics: *Revenue influenced by AI features, cost savings from AI automation, churn reduction attributable to AI, support volume change, time saved per user per week.*

The PM question: Is the AI investment generating measurable business outcomes, and can we attribute those outcomes to the AI specifically? This is the layer that boards and executives care most about, and the layer that is hardest to measure rigorously because attribution is genuinely difficult in complex product environments. The discipline of designing attribution approaches before launch, not after, is what separates teams that can answer these questions from those that cannot.

If you have metrics at only one or two of the three layers, you have a measurement gap that will eventually produce a conversation you are not prepared for. Build all three layers before you launch, even if the business metrics take longer to stabilize than the technical and product metrics.

· · ·

Evals: Building the Testing Infrastructure for AI Quality

Evals (evaluations) are the testing framework for AI systems. They are how you know, before a user encounters the output, whether the model is performing at the standard your product requires. Building a strong eval infrastructure is one of the highest-leverage investments an AI product team can make, because it is the foundation of every other quality improvement practice.

There are three categories of evals, each designed to catch different types of quality problems:

1. **Deterministic evals.**

   These test for correctness on questions that have clear right and wrong answers. Did the model follow the required output format? Did it include all required fields? Did it stay within specified constraints (length, language, content restrictions)? Did it correctly identify the category, entity, or attribute the task required? Deterministic evals should be automated and run on every model change. They are the regression tests of the AI product world.

2. **Probabilistic evals.**

   These test for quality on questions that require judgment: is this response helpful, accurate, and appropriate for the context? Is this summary faithful to the source document? Is this recommendation relevant to this user's actual situation? Probabilistic evals cannot be fully automated because the answers involve judgment. They can be partially automated using the LLM-as-judge pattern: a separate, carefully prompted model that evaluates the outputs of the product model against a defined rubric. Human evaluation remains important for calibration and for the highest-stakes quality questions, but LLM-as-judge dramatically reduces the cost of running probabilistic evals at scale.

3. Adversarial evals.

These test for robustness under conditions the model was not explicitly trained for: unusual inputs, edge cases, attempts to produce outputs the product should not produce, and combinations of inputs that surface unexpected behaviors. Adversarial evals are the most frequently skipped category and the most frequently regretted omission after launch. The failure modes that adversarial evals catch are almost never visible in normal usage until they are visible to a large number of users at once.

THE EVAL COVERAGE TEST

Before any AI feature launches, the team should be able to confirm the following: we have deterministic evals covering all format and constraint requirements. We have probabilistic evals covering the primary quality dimensions that matter to users. We have adversarial evals covering the failure modes we identified in pre-flight and design review. We have a plan for who runs these evals, on what cadence, and what happens when they fail. If any of these cannot be confirmed, the eval infrastructure is not ready.

. . .

AI Success Metrics by Product Type

The specific metrics that matter most vary significantly depending on the type of AI product you are building. A recommendation engine, a generative AI feature, and an agentic workflow have different quality dimensions, different failure modes, and different measurement approaches. The following frameworks are starting points, not exhaustive lists.

Recommendation and Personalization Systems

The primary quality dimension for recommendation systems is relevance: does the system surface the right content or options for the right user at the right time? Measure click-through rate and conversion rate on recommendations, but supplement these with downstream outcome metrics: did the user who acted on the recommendation achieve a positive outcome, or did they return, abandon, or reverse the action? Diversity metrics (is the system recommending from too narrow a slice of the available options?) and novelty metrics (is the system surfacing things the user would not have found on their own?) are often valuable additions that aggregate click-through rates systematically miss.

Generative AI Features

Generative AI features (writing assistants, summarization tools, code generation) require measurement approaches that capture both the quality of the output and the degree to which the output is actually used. Acceptance rate (what proportion of generated content is used without modification), modification rate (how much do users change the generated content before using it), and task completion rate (do users who engage with the generated content complete the underlying task more often than those who do not) together give a more complete picture than any single metric. Hallucination rate, measured through a combination of automated fact-checking and human review on a sample basis, should be monitored continuously for any generative feature that makes factual claims.

Agentic Products

Agentic products require a measurement framework that is sensitive to both the quality of individual actions and the quality of multi-step workflows. At the action level: accuracy rate, override rate, and irreversible error rate. At the workflow level: task completion rate, human intervention rate (how often does a human need to step in to correct or redirect the agent?), and time to completion compared to the non-AI baseline. Irreversible error rate deserves particular attention: even a small rate of irreversible errors can produce significant user harm at scale, and it should be monitored with a lower alert threshold than other error types.

. . .

Closing the Executive Perception Gap

Research conducted in 2026 found a striking disparity in how AI product performance is perceived across organizational levels: fifty-six percent of executives believe AI is embedded and working effectively across the product lifecycle, compared with only eighteen percent of managers who see the same reality on the ground. This gap is not primarily a communication problem. It is a measurement problem. When the metrics used to report AI performance to executives are disconnected from the metrics used to manage AI quality at the team level, the two groups develop fundamentally different pictures of the same product.

Closing this gap requires building a measurement framework that is **vertically coherent:** the business metrics that executives see should be traceable to the product metrics that PMs manage, which should be traceable to the technical metrics that engineers monitor. When a technical metric degrades, a PM should be able to predict which product metrics will follow and on what timeline. When a business metric improves, a PM should be able to identify which product and technical improvements drove it.

This vertical coherence does not happen automatically. It requires intentional design of the measurement architecture at the start, before the metrics are collected and before the stakeholder reporting is established. Teams that design measurement architecture after launch consistently find that the data they collected during launch is not structured in a way that allows the vertical connections to be made.

BUILDING A MEASUREMENT DASHBOARD YOUR CEO WILL TRUST

The most effective executive AI dashboards have three properties. First, they lead with business outcomes, not technical performance: revenue influenced, cost saved, time returned to users. Second, they show attribution clearly: this outcome changed by this amount and here is the evidence that the AI feature drove the change. Third, they include a health indicator that represents the PM's confidence in the quality and reliability of the AI, so executives understand that the business outcome metrics are sitting on a foundation that is being actively monitored. A dashboard that shows strong business outcomes without a quality health signal gives executives no way to know whether those outcomes are durable or fragile.

. . .

IN PRACTICE: THE METRICS THAT SAVED THE INVESTMENT CASE

We had built an AI feature that was performing well by the metrics we were tracking: high adoption, positive user feedback, strong engagement. When the annual planning cycle came around and we needed to defend the investment, we realized that none of our metrics answered the question the finance team was asking: what would have happened without the AI?

We had not built a control group. We had not designed the measurement architecture to support attribution. We had metrics that showed the product was good and no evidence that the AI was responsible for any specific portion of that goodness.

We spent six weeks after the planning cycle building a retrospective attribution model, using pre-launch behavioral data as a baseline and difference-in-differences analysis to isolate the AI's contribution. The result supported our case: the AI feature was responsible for a thirty-one percent reduction in the time users spent on the target task. But we had to fight for that finding because we had not designed for it. The conversation with the finance team was difficult and the outcome was closer than it should have been.

Since then, every AI feature I build starts with an explicit attribution design: what is the control condition, what is the measurement window, and what analysis approach will we use to isolate the AI's contribution? That design happens before the feature is built, not six weeks after the planning cycle when the data you need may no longer be collectable.

. . .

What Comes Next

With a measurement framework in place, you have the infrastructure to know whether your AI product is working. Chapter 9 covers the final stage of the execution layer: how to bring an AI product to market in a way that earns user trust from day one, navigates the regulatory and ethical considerations that are now unavoidable for any AI product in a serious market, and avoids the launch antipatterns that consistently destroy the trust you spent months building.

— · —

Continue to Chapter 9: Launching AI Products

CHAPTER NINE

Launching AI Products: Trust, Transparency, and Go-to-Market

How to bring an AI product to market in a way that earns user trust from day one, avoids the antipatterns that consistently destroy it, and positions the product for long-term competitive success.

In 2024, a well-funded AI startup launched a product that could genuinely do something remarkable: it could synthesize weeks of customer research into a structured insight report in under three minutes. The technology was real. The value proposition was clear. The launch failed. Not because users did not want what the product offered but because the launch communication had overpromised in ways that the product, impressive as it was, could not consistently deliver. Users arrived expecting magic and encountered something that was very good but not magic. The gap between expectation and experience was not large. But in an AI product, that gap is where trust lives and dies. The company spent the following year rebuilding a user relationship that a more carefully managed launch would never have damaged.

Launching an AI product is categorically different from launching a traditional software product, and the differences are not primarily technical. They are about the particular relationship between users and AI systems: the heightened expectations, the lower tolerance for confident errors, the specific anxieties about automation and displacement, and the trust dynamics that determine whether a user who encounters a failure gives the product another chance or abandons it permanently.

This chapter covers the launch layer end to end: how trust is built and how it is destroyed, the ethical and regulatory considerations that are now unavoidable for any AI product in a serious market, the go-to-market nuances that are specific to AI products, and the launch antipatterns that appear consistently across AI product failures and are almost entirely avoidable with the right preparation.

. . .

Trust as the Gate, Not the Goal

The most important reframe for AI product launches is this: trust is not a goal to be achieved after launch. It is a gate that must be cleared before launch. A user who encounters an AI product without a foundation of trust will interpret every error, every surprising output, and every moment of uncertainty as evidence that the product is not safe to rely on. A user who arrives with a foundation of trust will interpret the same experiences as evidence that the product is learning and improving. The same product, experienced through different trust frames, produces completely different adoption and retention outcomes.

Building the trust foundation before launch requires three things. **First, honest positioning:** the product should be described in launch communications in terms of what it reliably does, not what it impressively can do on its best day. The gap between reliable and impressive is the gap between a launch that builds trust and one that destroys it.

Second, **transparency about AI involvement:** users should know when they are interacting with an AI system, what the AI is doing on their behalf, and what constraints the AI is operating within. This is not just an ethical requirement in most jurisdictions. It is a trust-building practice. Users who understand what the AI is doing are more tolerant of its limitations than users who encounter AI behavior without context.

Third, **a visible and simple path to human support:** enterprise buyers in particular, and increasingly consumer users in high-stakes contexts, will not adopt an AI product that does not give them a clear way to reach a human when the AI is insufficient. This is not a sign of product weakness. It is a design feature that signals the product takes user needs seriously even when the AI cannot fully meet them.

> **The launch communication that earns trust describes what the product does reliably, acknowledges what it does not do, explains how users can tell the difference, and provides a clear path forward when the AI falls short. Any launch communication that cannot be written in those terms is a signal that the product is not ready to launch.**

. . .

Explainability in Practice: Showing Your Work Without Overwhelming Users

Explainability is one of the most discussed and most poorly implemented concepts in AI product design. The discussion usually conflates two very different things: technical explainability (understanding how the model produces an output, which is often genuinely difficult) and product explainability (giving users enough context about the AI's behavior to use the product confidently and appropriately, which is a design problem, not a technical one).

Product explainability does not require exposing the model's internals to users. It requires designing the user experience so that users can answer three questions without technical expertise: what did the AI do? Why did it do it, at a level of abstraction that makes sense to me? And what can I do if I think it got it wrong?

1. **What did the AI do?**

   This is the transparency layer: making the AI's actions and outputs visible and legible. Not buried in a log, not described in technical language, but surfaced in the user experience in plain language at the moment of relevance. "The AI has drafted a response based on your three most recent conversations with this customer." "This summary was generated from the five documents you selected." Simple, specific, and immediate.

2. **Why did it do it?**

   This is the provenance layer: giving users a human-readable explanation of the reasoning or inputs behind the AI's output. Not the technical mechanism but the product-level logic. "This recommendation is based on your purchase history and what similar users have found valuable." "This flagged item matched three of the five criteria you defined as high priority." The goal is not completeness. It is sufficiency: enough context that the user can evaluate the output intelligently.

3. What can I do if it got it wrong?

This is the agency layer: ensuring that the user always has a visible, simple path to override, correct, or dismiss the AI's output. The override should never be buried. It should never require more steps than accepting the output. And every override should be treated as a valuable signal, instrumented carefully, and fed back into the improvement cycle described in Chapters 6 and 8.

. . .

Ethical Guardrails as Product Requirements

The instinct in most product organizations is to treat ethical review as a final checkpoint before launch: a box to check, a conversation to have with legal and compliance, a set of constraints to layer onto a product that has already been designed. This instinct is expensive. Ethical problems discovered at the launch gate require redesign at the most costly possible moment, when the product is built and the team is under pressure to ship.

The alternative is to treat ethical requirements as first-class product requirements that shape the design from the beginning, with the same priority and visibility as performance or usability requirements. This is not a slowing-down of the process. It is a front-loading of decisions that will have to be made eventually, made at the moment when changing them is cheapest.

The ethical requirements that are most consistently underweighted in early AI product design fall into four categories:

1. **Bias and fairness.**

   Does the AI perform consistently across different user populations, demographic groups, and use cases? Inconsistent performance across groups is not just an ethical problem. It is a product quality problem and, in many jurisdictions, a regulatory problem. Bias evaluation should be part of the eval framework from Stage 4 of the lean AI build cycle, not a post-launch audit.

2. **Privacy and data use.**

   What user data does the AI system use, how is it stored, how long is it retained, and who has access to it? Users in most major markets now have legal rights around how their data is used in AI systems, and the regulatory landscape is tightening. Data use decisions made without explicit privacy design are product liability waiting to materialize.

3. **Scope and misuse.**

   What is the product designed to do, and what is it designed not to do? How will the product respond when users attempt to use it in ways that are outside its intended scope or that could cause harm? Scope definition and misuse prevention are product design decisions, not content moderation decisions. They should be designed into the system from the beginning, with explicit testing for the misuse patterns that are most likely given the product's capabilities and user base.

4. **Human oversight in consequential decisions.**

   For any AI product that influences consequential decisions (hiring, lending, medical triage, legal outcomes, financial advice), the design should ensure that a human is meaningfully in the loop at the decision point, not just notified after the fact. The regulatory requirements here are evolving rapidly and vary significantly by jurisdiction, but the product design principle is consistent: the higher the consequence of an error, the more explicit and substantive the human oversight should be.

. . .

Go-to-Market Nuances for AI Products

The go-to-market motion for an AI product differs from traditional software in ways that are easy to underestimate until they cause problems. Three nuances deserve explicit attention:

1. **Pricing for value, not for usage.**

   The instinct for AI products is often to price by usage: per query, per generation, per API call. This model is transparent and simple, but it creates a friction at the exact moment of value delivery that works against adoption. Users who are uncertain about value will use the product less to control cost, which reduces their exposure to value, which reinforces their uncertainty. The pricing models that drive the strongest AI product adoption are typically outcome-based or value-based: what does this product allow you to do that you could not do before, and how is the price calibrated to that outcome rather than to the cost of the inference?

2. **Positioning around the outcome, not the technology.**

   The most common AI product positioning mistake is leading with the technology: "powered by GPT," "built on our proprietary large language model," "uses advanced machine learning to..." These formulations place the AI at the center of the value proposition, which creates two problems. First, the technology is not defensible: competitors have access to the same models. Second, users do not buy technology. They buy outcomes. Position around what the user can now do, achieve, or avoid rather than around how the product works. The technology is interesting to engineers and investors. The outcome is what drives adoption.

3. **Managing expectation velocity.**

   AI products exist in a market where user expectations are being set by the most impressive AI demonstrations they have seen, not by the average performance of AI products in your category. This creates an expectation gap that is not your product's fault but is your product's problem to manage. The most effective approach is to anchor user expectations explicitly at onboarding: show users what the product does well, show them what it is not designed to do, and give them a calibrated sense of when they should trust the output and when they should verify it. Users with calibrated expectations are more satisfied and more retained than users with uncalibrated ones, even when the underlying product is identical.

· · ·

The Launch Antipatterns That Destroy Trust

These antipatterns appear with striking regularity across AI product launches. Each one is avoidable with the preparation this chapter describes. Each one, when it occurs, produces trust damage that is disproportionate to the scale of the failure.

Antipattern 1: Overclaiming at launch.

The launch communication describes capabilities that the product cannot consistently deliver. Users arrive with expectations calibrated to the marketing and encounter a product calibrated to reality. The gap, even when small, produces a trust deficit that the product spends months trying to close.

The fix: Write the launch communication after reviewing the eval results, not before. Every capability claim should be traceable to a specific, measured eval outcome. If the eval does not support the claim, the claim should not be in the launch communication.

Antipattern 2: Launching to everyone before anyone.

The product goes to the full user population before the team has evidence of how it performs with real users in real conditions. Problems that would have been caught in a controlled pilot surface at scale, where they affect large numbers of users simultaneously and generate negative feedback loops that are much harder to reverse.

The fix: Pilot launches are not optional for AI products. The pilot population should be large enough to surface real-world failure modes and small enough that failures are manageable. Five to ten percent of the target population is a reasonable starting point. Full launch should not happen until the pilot metrics are stable and healthy.

Antipattern 3: No visible AI disclosure.

Users discover they have been interacting with an AI system through inference rather than explicit disclosure. In consumer products, this generates a sense of deception that is disproportionate to the actual harm. In regulated industries, it may generate legal liability. In both cases, it destroys the trust that the product needs to retain users.

The fix: AI disclosure should be visible at the point of first AI interaction, not buried in terms of service. The format and specificity of the disclosure should be calibrated to the consequence of the AI's actions: low-stakes interactions warrant a simple indicator; high-stakes interactions warrant explicit explanation.

Antipattern 4: **Treating the launch as the finish line.**

The team treats full launch as the end of the build cycle rather than the beginning of the learning cycle. Monitoring is set up but not actively reviewed. The next sprint begins before the launch metrics have stabilized. Silent degradation begins and is not caught because no one is looking at the right signals.

The fix: Define the post-launch monitoring plan before launch, not after. Assign explicit ownership for each metric category. Schedule the first post-launch review for one week after full launch, regardless of what the metrics look like at that point. The discipline of reviewing early and often is what separates teams that catch problems at two weeks from teams that catch them at two months.

Antipattern 5: **Ignoring the regulatory environment.**

The product launches without an explicit accounting of the regulatory requirements that apply to AI products in the relevant jurisdictions and use cases. As AI regulation has accelerated in the EU, UK, US, and major markets across Asia, the cost of discovering regulatory non-compliance after launch has grown from a legal inconvenience to a genuine product existential risk.

The fix: Regulatory review should begin at the product design stage, not at the launch gate. The specific requirements vary by jurisdiction, industry, and use case, but the general principle is consistent: the higher the consequence of an error and the more personal data involved, the more stringent the regulatory requirements will be. Engage legal and compliance as product partners, not as launch checkboxes.

THE RESPONSIBLE AI LAUNCH CHECKLIST

Before any AI product goes to full launch: launch communication has been reviewed against eval results and every claim is supported. AI disclosure is visible and appropriate to the consequence of the AI's actions. Pilot results are stable and healthy across all three metric layers. Ethical requirements (bias, privacy, scope, oversight) have been reviewed and addressed. Regulatory requirements have been identified and satisfied. Post-launch monitoring ownership is assigned and the first review is scheduled. The path to human support is visible and functional. If any item on this list cannot be confirmed, the launch should wait.

· · ·

IN PRACTICE: WHAT A RESPONSIBLE AI LAUNCH ACTUALLY LOOKS LIKE

The most successful AI product launch I have been part of was also the most uncomfortable to lead in the weeks before it happened. We had a strong product, strong eval results, and a pilot that had gone well. We also had a legal review that identified a disclosure requirement we had not designed for, a bias eval that showed meaningfully lower performance for one user segment we had underrepresented in our test data, and a pilot feedback pattern that suggested our onboarding was setting expectations too high for a specific set of use cases.

Any one of those findings, in a different culture, would have been minimized and moved past in the push to hit the launch date. In this case, we delayed by three weeks. We fixed the disclosure design. We expanded the training data for the underperforming segment and re-ran the bias eval until the performance gap was within acceptable bounds. We redesigned the onboarding for the problematic use cases to set more accurate expectations.

The launch, when it happened, was quieter than a launch without those delays would have been. We did not have a dramatic day-one story. What we had was a week-one retention rate that was thirty percent higher than our category benchmark, a month-one NPS that our investor called the best he had seen at this stage, and zero regulatory inquiries in the first year of operation. The three weeks we lost at the launch gate came back to us many times over in the first year of growth. A responsible launch is not a slower launch. It is a better launch.

. . .

Closing Part Three

The three chapters in Part Three gave you the complete execution layer of the AI PM toolkit. Chapter 7 gave you the end-to-end process for scoping and building an AI feature correctly. Chapter 8 gave you the measurement framework for knowing whether it is working. This chapter gave you the launch framework for bringing it to market in a way that earns and sustains user trust.

Part Four turns to the dimension that no competitor has addressed: your career. Chapter 10 gives mid-level PMs the specific, tactical guidance for navigating the AI PM career split, closing the salary gap, and building the portfolio and positioning that opens the right doors. Chapter 11 gives senior leaders the framework for building AI product organizations that are genuinely capable, not just AI-aspirational.

— · —

Continue to Part Four: Win the Career · Chapter 10: The AI PM Career Playbook

PART FOUR: WIN THE CAREER

CHAPTER TEN

The AI PM Career Playbook

How to choose your specialization, build the portfolio that signals genuine capability, close the salary gap, and position yourself for the AI PM roles that are redefining what the career looks like.

In 2025, a product manager with seven years of experience and a strong track record at two well-regarded companies applied for an AI PM role at a major technology company. She did not get an interview. Three months later, a PM with four years of experience at a less prominent company got the same role at a higher level. The difference was not pedigree or years of service. It was that the second PM had spent eighteen months deliberately building the specific portfolio, the specific language, and the specific positioning that the AI PM hiring market now recognizes as genuine capability. The first PM had done excellent work that the traditional PM hiring market valued highly. The AI PM hiring market is a different market, and it selects for a different kind of evidence.

The salary data established at the beginning of this book describes a gap that is real and growing. But the more important insight is not the size of the gap. It is what it implies about how the AI PM career works: it is not a linear extension of the traditional PM career. It is a different track with different selection criteria, different portfolio requirements, and a different set of conversations you need to be able to have to get the roles that pay at the top of the range.

This chapter is the tactical guide to navigating that track: which specialization to pursue, how to build the portfolio evidence, how to position and negotiate for the compensation that reflects your capability, and how to prepare for the AI PM interview in a way that goes well beyond the standard PM interview preparation.

. . .

The Four AI PM Specializations

The AI PM role is not a single job. It is a family of roles organized around different relationships between product management and AI capability. Understanding which specialization fits your background, interests, and target company is the first strategic decision in the AI PM career.

AI Product Strategy PM

Owns the AI product vision, strategy, and roadmap at the business unit or company level. The primary job is translating AI capability into competitive product strategy: where do we play, how do we win, and how does AI create defensible advantage in those choices? This role exists primarily at senior and principal PM levels and requires a combination of deep market understanding, strong executive communication, and the strategic AI fluency that Parts One and Two of this book are designed to build.

Signal work for your portfolio: A documented case where you drove an AI product strategy decision that changed the company's direction, with measurable business outcomes.

Who is hiring for this: Late-stage startups building AI-native products, large technology companies with dedicated AI product organizations, and enterprise software companies undergoing AI transformation.

AI Feature PM

Owns the end-to-end development and performance of AI features within a product. This is the most common AI PM role and the most accessible entry point for mid-level PMs making the transition. The primary job is applying the frameworks in Parts Two and Three of this book: scoping AI features correctly, managing the build cycle, measuring quality rigorously, and launching in a way that earns user trust.

Signal work for your portfolio: A shipped AI feature with documented eval results, user adoption metrics, and a clear attribution of business impact.

Who is hiring for this: Almost every technology company building AI products at any scale. This is the highest-volume AI PM role in the market.

AI Platform PM

Owns the internal AI infrastructure and tooling that other product teams build on. The customer is internal: ML engineers, data scientists, and feature PMs who need reliable, well-documented AI platform capabilities to do their work. This role requires stronger technical fluency than the other specializations and a particular facility for understanding developer experience and platform economics.

Signal work for your portfolio: Evidence of translating technical platform capabilities into product team productivity improvements, with specific metrics.

Who is hiring for this: Large technology companies with internal AI platforms, companies building AI developer tools, and enterprises that are centralizing AI infrastructure across business units.

AI Safety and Responsible AI PM

Owns the governance, safety, and ethical dimensions of AI product development. This specialization has emerged rapidly as regulatory pressure has increased and as the consequences of AI product failures have become more visible and more costly. The primary job is the work described in Chapters 5 and 9: failure mode analysis, ethical requirement design, regulatory compliance, and the organizational processes that make responsible AI a consistent practice rather than a periodic audit.

Signal work for your portfolio: A case where you identified a safety or ethical risk in an AI product before launch and drove a design change that addressed it, with documentation of the risk, the decision, and the outcome.

Who is hiring for this: Large technology companies with dedicated responsible AI teams, companies operating in regulated industries (financial services, healthcare, legal), and any company operating in the EU under the AI Act.

CHOOSING YOUR SPECIALIZATION

The right specialization is the intersection of three things: where your existing experience gives you credible foundation, where the work genuinely interests you enough to go deep, and where the market is currently paying a premium. All four specializations are well-compensated relative to the traditional PM market. The AI Feature PM role is the most accessible transition. The AI Strategy PM role has the highest ceiling. The Responsible AI PM role has the fastest-growing demand. The Platform PM role has the most stable long-term positioning because platform capability is the foundation everything else builds on.

. . .

Building the Portfolio That Signals Genuine Capability

The AI PM hiring market has a specific problem: it is very easy to claim AI PM experience and very hard to verify it. Every PM who approved a feature that used an AI API can describe themselves as having AI product experience. Every PM who attended an AI conference can describe themselves as having AI product knowledge. The signal-to-noise ratio in AI PM resumes and LinkedIn profiles is low, and experienced hiring managers know it.

The portfolio that breaks through this noise is not a longer list of AI tools you have used or AI projects you have been adjacent to. It is a small number of specific, documented examples of AI product decisions you drove, with outcomes that demonstrate the kind of thinking this book has been developing.

1. **One documented build.**

   A specific AI feature or product you were the primary PM for, from problem identification through launch and iteration. The documentation should cover: the user problem and the evidence for it, why AI was the right solution, the key design decisions and their rationale, the eval framework and results, and the measured business outcome. This does not need to be a large or complex feature. It needs to be one where you can walk through every decision with the specificity and confidence that only comes from genuine ownership.

2. **One documented failure and recovery.**

   The AI PM hiring market, at the senior level especially, is deeply skeptical of candidates who only describe successes. A documented case where an AI feature underperformed, where you identified the root cause through rigorous measurement, and where you drove the iteration that produced improvement, is more persuasive evidence of genuine AI PM capability than three success stories told without friction.

3. One demonstrated strategic contribution.

An example where your AI product thinking influenced a decision above the feature level: a strategy choice, an organizational design, a build-versus-buy decision, a market positioning call. This signals that you are not just an execution PM but a PM whose AI fluency generates value at the strategic level.

4. A visible point of view.

A written or published perspective on an AI product topic that demonstrates independent thinking: a blog post, a LinkedIn article, a talk at a meetup or conference, a contribution to a relevant community. The content matters less than the evidence that you have a perspective you are willing to articulate publicly and defend in conversation. Hiring managers for senior AI PM roles consistently report that candidates with a visible point of view advance further in the process than equally qualified candidates without one.

. . .

Closing the Salary Gap: Positioning and Negotiation

The salary gap between AI PMs and traditional PMs is not a market inefficiency that will self-correct. It reflects a genuine difference in the scarcity of the capability, the leverage that capability creates for the hiring company, and the competitive pressure among employers for a limited pool of candidates. Closing it requires positioning and negotiation practices that are different from those that work in the traditional PM market.

The most important positioning shift is **leading with outcomes, not with credentials.** Traditional PM compensation conversations often center on years of experience, company pedigree, and scope of previous roles. AI PM compensation conversations that close at the top of the range center on specific, measurable outcomes: the revenue influenced, the cost reduced, the quality improvement achieved through AI products you owned. If you have built the portfolio described in the previous section, you have the raw material for this conversation. The work is translating that material into the language of business outcomes.

The negotiation dynamic for AI PM roles also differs from traditional PM roles in one important way: the employer's outside option is worse. A company that cannot hire a strong AI PM does not simply hire a slightly weaker AI PM. It delays or degrades its AI product work, which has compounding consequences in a fast-moving market. This asymmetry gives well-positioned AI PM candidates more negotiating leverage than they typically recognize, and it is leverage that is lost by candidates who do not name it.

THE THREE NEGOTIATION ANCHORS FOR AI PM ROLES

First, anchor to the market rate for the specialization, not the general PM market. The $245,000 average for AI-focused PMs is a reference point, not a ceiling, and the range within specializations varies significantly. Know the specific range for your specialization and level before any compensation conversation.

Second, anchor to the value your specific experience creates for this company. What AI product problems does this company have that your portfolio proves you can solve? The more specifically you can connect your capability to their actual challenges, the more the conversation shifts from what the market pays to what your contribution is worth to them.

Third, anchor to the total compensation package, not just base salary. Equity, bonuses tied to AI product outcomes, and professional development allowances for conference attendance, research access, and continued learning are all negotiable components that AI PM candidates at strong companies can often secure with a direct ask.

. . .

What the Major Technology Companies Are Actually Hiring For

The AI PM job descriptions posted by major technology companies are often written to be comprehensive rather than precise, which makes them less useful as signals of what the hiring team actually cares about. The following observations are drawn from patterns across hiring conversations and role analysis at the companies that are currently defining what AI PM excellence looks like.

Google DeepMind and Google's AI product teams, **weight technical fluency and research collaboration most heavily.** The ability to work effectively with research teams, to translate research capabilities into product hypotheses, and to design evals that meet the rigor of a research environment is more valued than broad product execution experience.

Amazon's AI product roles **weight business outcome ownership and data rigor most heavily.** The expectation is that an AI PM can define the business case, own the measurement infrastructure, and defend the

ROI with the specificity that Amazon's leadership principles demand. The 'working backwards' discipline, applied to AI product problems, is the core skill the hiring process is designed to evaluate.

Microsoft's AI product roles **weight enterprise go-to-market and responsible AI most heavily.** Given Microsoft's enterprise customer base and its public commitments around responsible AI, the ability to navigate enterprise adoption dynamics, regulatory considerations, and trust-building at the organizational level is more differentiating than technical AI fluency.

Meta's AI product roles **weight scale, experimentation, and user behavior analysis most heavily.** The ability to design and interpret experiments at massive scale, to understand user behavior signals in AI product contexts, and to operate effectively in a data-driven culture where every product decision is expected to be grounded in evidence is the core differentiator in their hiring process.

· · ·

The AI PM Interview: How to Stand Out

The AI PM interview process at senior companies has evolved significantly from the traditional PM interview format. The product sense questions still exist, but they are now supplemented with AI-specific dimensions that test for the kind of thinking this book has been developing.

1. **The AI product design question.**

   You will be asked to design an AI feature or product for a specific use case. The interviewer is not primarily evaluating whether your solution is technically feasible. They are evaluating whether you start with the user problem, whether you ask the right questions about data and failure modes, whether you define success in measurable terms, and whether your design reflects the trust and transparency principles that distinguish thoughtful AI product work from feature-first thinking.

2. **The failure mode question.**

You will be asked about a time when an AI product or feature you worked on did not perform as expected. The interviewer is evaluating your diagnostic rigor: did you identify the root cause precisely? Did you distinguish between a model problem, a data problem, a design problem, and a measurement problem? And did you drive a response that addressed the actual cause rather than the visible symptom? Candidates who describe AI failures as technical problems they handed to engineering consistently underperform candidates who describe them as product problems they owned through to resolution.

3. **The metrics and measurement question.**

You will be asked how you measure the success of an AI feature. The answer the interviewer is looking for goes beyond adoption and engagement. It demonstrates awareness of the three-layer metrics stack, understanding of the difference between deterministic and probabilistic quality, and a specific approach to attribution that would hold up in a conversation with a skeptical finance team.

4. **The strategic question.**

You will be asked how AI changes the strategy for a specific product or market. The answer the interviewer is looking for demonstrates the kind of thinking developed in Chapters 3 and 4: the ability to map AI capability to competitive dynamics, to identify where AI creates defensible advantage versus commodity capability, and to articulate a strategic response that is specific to the context rather than generically 'AI-first.'

IN PRACTICE: HOW I MADE THE TRANSITION AND WHAT I WOULD DO DIFFERENTLY

I made the transition to a dedicated AI PM role after four years as a generalist PM. The transition took longer than it needed to because I made the mistake that most transitioning PMs make: I tried to demonstrate AI PM capability through education rather than through evidence. I took courses. I got certifications. I read research papers. None of it moved the needle in hiring conversations, because none of it answered the question the hiring manager was actually asking: have you done this before, and can you show me the work?

The shift happened when I stopped focusing on credentials and started focusing on one thing: getting a single AI feature shipped at my current company, owning it completely, documenting it rigorously, and being able to walk through every decision in detail. That feature was not large. It was not technically sophisticated. But I knew it better than any candidate the hiring team had seen, and that depth of ownership was the thing that differentiated me.

What I would do differently is start the documentation earlier. By the time I was preparing for interviews, some of the early decision context had faded, and the eval data that would have been most compelling required reconstruction from memory. Document every AI product decision as you make it, not in retrospect. The portfolio is not something you build before a job search. It is something you build every day while doing the work. The job search is just the moment when you surface it.

. . .

What Comes Next

Chapter 11 closes the career section from the other direction: for the senior leaders and CPOs who are not navigating the job market but building the organizations that AI PM talent joins. How do you structure an AI product organization, hire for genuine AI PM capability rather than impressive credentials, and build a culture where AI product work compounds over time rather than cycling through the same learning curve with every new team member?

— · —

Continue to Chapter 11: Leading AI Product Teams

CHAPTER ELEVEN

Leading AI Product Teams: Culture, Talent, and the Future

How to build, structure, and scale an AI product organization that is genuinely capable, closes the perception gap between leadership and execution, and creates the compounding culture that makes AI product work better over time.

A CPO at a well-regarded enterprise software company was asked at a board meeting how the company's AI product capability compared to its competitors. She gave a confident answer: strong, improving, ahead of the category average. The board was satisfied. Three weeks later, she sat in a product review with her senior PMs and heard a different story: a team stretched across too many AI initiatives, an eval infrastructure that existed in one team and not others, a pattern of AI feature launches that looked good in the demo and underperformed in production, and a growing sense among the team that the company's AI ambitions were outrunning its AI execution capability. The gap between the answer she gave the board and what she heard in that review was not dishonesty. It was the organizational equivalent of the fifty-six percent problem: leadership and execution were operating with different pictures of the same reality, and neither picture was complete.

Building a genuinely capable AI product organization is the hardest leadership challenge in the product management profession right now. Not because the technical problems are unsolvable but because the organizational problems are underappreciated. Most product leaders are focused on hiring AI PMs, buying AI tools, and announcing AI initiatives. The ones building durable capability are focused on something harder: creating the conditions under which AI product work compounds over time rather than cycling through the same learning curve with every new initiative.

This chapter is for those leaders. It covers organizational design, talent strategy, the culture that makes AI product work sustainable, and the forward view on where the AI PM role is heading so you can build an organization that is positioned for what comes next, not just what exists now.

. . .

The Organizational Design Question: Structure That Enables, Not Constrains

The most common organizational design mistake in AI product teams is treating AI as a capability that needs its own separate structure. A centralized AI team, a dedicated AI center of excellence, a VP of AI Products reporting to the CEO: these structures are appealing because they create visible AI investment and a clear accountability for AI outcomes. They are problematic because they separate AI capability from the product context where it creates value, producing technically sophisticated AI work that is systematically disconnected from user problems.

The organizational models that produce the best AI product outcomes are those that embed AI capability within product teams while providing shared infrastructure and governance at the portfolio level. The following three models represent the most common effective approaches, each with different trade-offs depending on company size, AI maturity, and the concentration of AI work across the product portfolio.

The Embedded Model

AI capability is distributed across product teams. Each team has at least one PM with strong AI fluency, paired with ML engineers and data scientists embedded in the team. AI governance, shared infrastructure, and portfolio-level coordination happen through a lightweight steering function, not a separate organizational unit.

Best for: Companies where AI capability needs to be close to the product context, where the AI work is distributed across multiple product areas, and where the organizational culture supports distributed accountability.

Watch for: Without strong governance mechanisms, this model produces shadow AI proliferation: each team builds its own AI infrastructure, creating redundancy, inconsistency, and a governance gap that becomes expensive to close at scale.

The Hub-and-Spoke Model

A central AI platform team owns the shared infrastructure, evaluation frameworks, and governance standards. Product teams own the AI features and user experience. The platform team acts as an internal service provider and standards body, not as the builder of product-facing AI features.

Best for: Companies that have reached a scale where AI infrastructure efficiency matters, where regulatory or compliance requirements demand consistent governance, and where the AI platform itself has become a product that other teams depend on.

Watch for: The platform team can become a bottleneck that slows product team velocity if its governance function is designed as a gate rather than an enabler. The CPO's job is to ensure the platform team is measured on product team's velocity, not just on technical quality.

The Accelerator Model

Small, dedicated cross-functional AI squads (two to four people from product, design, and engineering) are formed for specific high-priority AI initiatives, operate with significant autonomy and speed for a defined period, and then either scale into the core product organization or sunset. This model, referenced in Chapter 4, is most effective for organizations that need to move fast on a specific AI opportunity without restructuring the entire product organization.

Best for: Companies making a significant AI bet in a specific area, organizations that need to demonstrate AI progress quickly, and situations where the standard product development process is too slow for the opportunity window.

Watch for: Accelerator squads can create two-tier organizations where the accelerator team is seen as the exciting work and the core team is seen as the maintenance work. Managing this perception gap is a leadership responsibility, not an HR problem.

There is no universally correct organizational model for AI product teams. The right model depends on your company's size, AI maturity, and product structure. What is universally correct is that the model should be chosen deliberately, with explicit attention to the failure modes each model creates, rather than allowed to emerge organically from a series of individual hiring decisions.

. . .

Hiring AI PMs: What Actually Signals Capability

The AI PM hiring market has a high noise-to-signal ratio, as established in Chapter 10. For hiring managers, this creates a specific challenge: the standard PM hiring signals (pedigree, years of experience, scope of previous roles) are poor predictors of AI PM performance, and the AI-specific signals that do predict performance are not reliably visible in a standard resume or interview process.

The following signals have proven to be the most reliable predictors of AI PM performance across hiring processes at companies that have had time to validate their predictions with actual outcomes:

1. **Specific, documentable AI product ownership.**
   Not involvement in AI projects. Ownership: the PM made the key decisions, defined the success criteria, drove the evaluation, and can walk through every significant choice with the specificity that only comes from genuine accountability. The interview question is not 'tell me about an AI product you worked on' but 'walk me through the most important product decision you made in your last AI feature and why you made it.' The depth and specificity of the answer separates owners from adjacents.

2. **Failure mode fluency.**
   The ability to name the specific ways an AI feature might fail, distinguish between different categories of failure, and describe a design response to each one is a strong signal of genuine AI product experience. PMs who have built AI products in production have encountered failures that PMs who have only designed AI features have not. This experience is difficult to

simulate and easy to detect in a well-structured interview conversation.

3. Measurement sophistication.

The ability to describe how they would measure an AI feature before they build it, including the attribution approach and the eval framework, is a reliable signal of the analytical discipline that AI product work requires. PMs who describe AI success in terms of adoption and NPS alone have not yet developed the measurement sophistication that the most demanding AI product environments require.

4. Cross-functional credibility.

AI PM performance is highly dependent on the quality of the relationship with ML engineers and data scientists. In the interview process, ask how the candidate has navigated disagreements with technical teams on AI product decisions. PMs who describe these relationships as smooth and collaborative without tension are either working in unusually frictionless environments or are not yet aware of where the friction is. PMs who can describe specific disagreements, the substance of the disagreement, and how they resolved it have the cross-functional experience that the role requires.

WHAT NOT TO WEIGHT IN AI PM HIRING

AI certifications and online course completions are weak signals. They indicate interest and effort, not capability. Weight them the way you would weight any other credential: as evidence of motivation, not as evidence of competence.

Familiarity with specific AI tools is also a weak signal. The tools change faster than hiring cycles. A candidate who has learned five AI tools and can explain what problem each one solves and what its limitations are is more valuable than a candidate who has used fifteen tools and cannot articulate the trade-offs between them.

Brand pedigree (having worked at a well-known AI company) is a moderate signal at best. The AI PM role varies enormously across companies, and a PM who was a strong contributor at a company with excellent AI infrastructure may struggle in an environment where they have to build that infrastructure themselves.

. . .

Closing the 56 Percent Versus 18 Percent Gap

The gap between executive perception and manager-level reality on AI product maturity is not a communication problem that better reporting will solve. It is a structural problem: executives and managers are looking at different evidence, using different standards of what counts as embedded AI capability, and operating in different feedback loops that reinforce their respective pictures of the same organization.

Closing the gap requires three things from senior product leaders. **First, a shared definition of what AI capability actually means.** Not 'we have AI features in production' (too low a bar) and not 'we have a world-class AI organization' (too vague to measure). A specific, observable definition: every product team has at least one PM with demonstrated AI product ownership, every AI feature has an eval framework before it ships, every AI product decision is connected to a measurable business outcome. When the definition is specific, the gap between executive and manager perception becomes a diagnostic, not a dispute.

Second, **regular calibration between leadership and execution on AI product quality.** Not a quarterly business review where the AI dashboard shows green. A regular, structured conversation where product managers present the failure modes they are managing, the assumptions they are testing, and the places where the AI is underperforming against its design goals. Leaders who only see the green dashboards are the ones who give confident answers to board questions that their team would not recognize.

Third, **a reward structure that makes quality visible.** If the organization rewards shipping and does not reward the rigorous quality work that makes AI products trustworthy over time, it will get shipping. The CPO who wants to close the perception gap needs to make the quality work visible in performance conversations, in promotion decisions, and in the organizational stories that define what excellent AI product work looks like.

. . .

Building an AI Product Culture That Lasts

Culture in AI product organizations is shaped by the same forces that shape culture in any organization: what gets rewarded, what gets tolerated, and what the leaders model. But AI product culture has specific dimensions that require explicit attention because they do not emerge naturally from the standard product management operating model.

1. **A culture of honest measurement.**

    AI products produce a lot of metrics, and metrics are easy to select for flattery. The culture the best AI product organizations have built is one where the most important metrics are the ones that are hardest to look good on, and where presenting an honest picture of where the AI is underperforming is valued more than presenting an optimistic picture of where it is succeeding. This culture does not emerge from a values statement. It emerges from leaders who model it: who share their own team's failure data openly, who ask hard questions about the metrics being presented, and who make it safe to bring problems to the surface rather than managing them quietly until they become crises.

2. A culture of continuous learning.

The learning operating system described in Chapter 6 is not just a set of practices. It is a cultural posture: the belief that the organization's most valuable resource is its ability to learn faster than the market, and that learning is therefore not a side activity but a primary organizational responsibility. Leaders who model this posture ask what the team learned from a launch before asking what the metrics look like. They invest in the rituals and tooling that make learning systematic. And they treat the question 'what were we wrong about?' as a sign of organizational health, not organizational failure.

3. A culture of user-centered AI.

The most insidious failure mode in AI product culture is the gradual drift from user-centered thinking to capability-centered thinking: building AI features because the capability exists, because the team finds it interesting, or because the market expects AI investment, rather than because it solves a problem users actually have. The culture that prevents this drift is one where every AI product decision is anchored to a specific user problem with specific evidence, where the pre-flight checklist is a cultural norm rather than a formal process, and where the question 'what problem does this solve for which users?' is asked and genuinely answered before any AI initiative receives investment.

4. **A culture of responsible innovation.**

AI products can cause harm at a scale and speed that traditional software cannot, and the organizations that have built lasting reputations in AI product are those that treat responsible innovation as a competitive advantage rather than a constraint. This culture is built by leaders who include ethical requirements in the product definition from the beginning, who make the responsible AI launch checklist a standard part of the process, and who create the psychological safety for team members to raise ethical concerns without those concerns being treated as obstacles to shipping.

. . .

Future-Proofing Your Team for the Next Wave

The AI landscape in 2026 is not the end state. It is a waypoint in a longer transition whose direction is clear even if its pace is uncertain. The organizations that will be best positioned in 2028 and beyond are not those that have optimized for the current state of AI capability but those that have built the organizational capacity to absorb and apply new AI capabilities as they arrive.

Three capabilities are most important for future-proofing an AI product organization. **First, deep user understanding that is independent of any specific AI capability.** The teams that are most effective at applying new AI capabilities are those that have an exceptionally clear picture of user problems, workflows, and unmet needs. When a new AI capability arrives, these teams can immediately identify which user problems it addresses and how it needs to be designed to fit into real user behavior. Teams that are primarily capability-driven have to rediscover the user problem with each new capability, which is slower, more expensive, and more likely to produce features that solve the capability's design problem rather than the user's actual problem.

Second, **evaluation infrastructure that can assess any AI capability against your quality standards.** Teams that have built a strong eval culture and a flexible evaluation infrastructure can assess a new AI capability against their user quality requirements in days rather than months. Teams without that infrastructure have to build it from scratch every time they encounter a new capability, which creates a consistent lag between capability availability and product deployment.

Third, **a learning operating system that treats organizational knowledge as a compounding asset.** The organizations that compound their AI product capability over time are those where every launch, every failure, and every user insight is systematically captured, processed, and fed back into the next cycle. The learning does not evaporate when team members leave. It does not have to be rediscovered with each new initiative. It accumulates in the organization's processes, documentation, and shared mental models in a way that makes each successive AI product better than the last.

IN PRACTICE: BUILDING THE TEAM I WISH I HAD JOINED

When I had the opportunity to build an AI product team from scratch, I made a conscious decision to prioritize three things that I had not seen prioritized in any of the AI product organizations I had been part of before.

The first was eval infrastructure before features. Before we built a single AI feature, we built the evaluation framework that would be used to assess every AI feature we shipped. It took six weeks and felt like a delay. In the following year, it saved us from shipping four features that would have underperformed and gave us the measurement foundation to improve three features that launched below target. The six weeks came back many times over.

The second was psychological safety around failure. In the first month, I shared a detailed post-mortem on an AI feature I had shipped at a previous company that had failed, including my specific mistakes in the scoping, the eval design, and the launch. The effect on the team's willingness to surface problems early was immediate and visible. Leaders who model vulnerability about failure create the conditions for team members to raise concerns before they become crises.

The third was a no-AI-feature-without-a-user-problem rule. Every AI feature proposal had to start with a specific user problem, supported by specific evidence. Not a technology capability looking for a use case. A user problem looking for the best solution, which might or might not be AI. This rule was resisted at first by team members who were excited about AI capabilities they wanted to build. Within six months, it was the cultural norm that defined how the team thought about its work. The features we shipped as a result of this discipline had higher adoption, higher retention, and higher user trust than anything I had seen built under the capability-first model. That is the organization I wish I had joined earlier in my career. Building it was the best product management work I have ever done.

Closing Part Four

The two chapters in Part Four completed the career layer of the book. Chapter 10 gave individual contributors the specific tactical guidance for navigating the AI PM career: the specializations, the portfolio, the salary negotiation, and the interview preparation. This chapter gave senior leaders the organizational and cultural framework for building AI product teams that compound over time.

The conclusion that follows brings the full book together into a single integrated action plan. It is short by design. Everything you need to act has been in the chapters that preceded it. The conclusion's job is to make the first step clear, the next step visible, and the series of steps that follows feel achievable from wherever you are starting today.

Continue to Conclusion: Your AI Advantage Starts Now

CONCLUSION

Your AI Advantage Starts Now

You have covered a lot of ground. Eleven chapters, four parts, a complete arc from the foundational mental models through strategy, execution, and career. The risk at the end of a book like this is the same risk at the end of any intensive learning experience: leaving with a clear sense of what you know and an unclear sense of what to do first.

This conclusion is designed to solve that problem. It is not a summary of what you have read. It is a bridge from reading to doing, organized around the thirty days immediately following this page.

. . .

What You Now Have That Most PMs Do Not

Before the action plan, a brief accounting of what this book has given you, because understanding what you have is part of knowing how to use it.

You have a mental model of the AI product landscape that is current, specific, and oriented around competitive dynamics rather than technology taxonomy. You can map any AI product to its tier, identify its defensibility, and articulate what the PM's job actually is at that tier. Most of your peers cannot do this consistently.

You have a strategy framework that is designed for the rate of change AI markets actually operate at, not for the stable environment that traditional roadmap planning assumes. You have product principles that generate real decisions, an adaptive planning cycle that updates gracefully when the market moves, and a learning operating system that turns every experiment into compounding advantage.

You have an execution playbook that covers the full arc from pre-flight to pilot to launch to iteration, with specific frameworks for the failure modes, measurement gaps, and trust dynamics that make AI product work different from traditional product work. You know how to design for agentic AI. You know how to measure what actually matters. You know how to launch in a way that earns trust rather than spending it.

And you have a career framework that is specific enough to act on: a specialization to pursue, a portfolio to build, a salary conversation to have, and an interview to prepare for with the evidence and language the AI PM hiring market actually responds to.

> **The gap between where you were before page one and where you are now is real. The gap between where you are now and where the AI PM market rewards you for being is a function of one thing: what you do in the next thirty days.**

. . .

The 30-Day Action Plan

The plan is organized into four weeks, each with a primary focus and a specific deliverable. The deliverables are not exercises. They are real work products that will serve you in your current role and in every career conversation you have from this point forward.

Week 1 Audit and orient.

Run the three diagnostic questions from Chapter 3 on your current product or the product you are closest to. Which tier is it in? How explainable is it to AI-mediated discovery systems? Does it generate a data flywheel? Write the answers down, specifically enough that you could share them with your team. The audit itself will surface the strategic conversations your product organization most needs to have.

Week 2 Write your product principles.

Draft three to five product principles for your current AI product or AI initiative, using the vague-versus-useful framework from Chapter 4. Test each one against a real decision your team has faced in the last month. If the principle would not have changed or clarified that decision, rewrite it until it would. Share the principles with one trusted colleague and ask them to try to break them. The friction this creates is the point.

Week 3 Build one portfolio piece.

Identify the AI feature or product decision in your recent work that you know most thoroughly. Document it using the portfolio framework from Chapter 10: the user problem and evidence, the key decisions and rationale, the eval results, and the measured business outcome. If the documentation surfaces gaps in what you measured or recorded, those gaps are the most important thing to fix in your next AI feature. Start documenting from the beginning, not in retrospect.

Week 4 **Start the conversation.**

Have one conversation that would not have happened before you read this book. With your manager about the AI PM specialization you are pursuing and what support you need. With your engineering lead about the failure modes in your current AI feature that you have not yet designed against. With a peer PM about the measurement gap in your team's AI product metrics. The conversation does not need to be large. It needs to be the one that has not happened yet and that your new understanding makes possible.

USING THE AI ADVANTAGE TRAINING APP ALONGSIDE THIS BOOK

Each chapter in this book maps to a module in the AI Advantage training app. The app is where the frameworks become fluent: scenario-based practice for the pre-flight checklist, interactive eval design exercises, interview preparation for each of the four AI PM specializations, and a portfolio builder that guides you through the documentation process with structured prompts and peer feedback. The book builds the mental model. The app builds the muscle memory. They are designed to be used together, and the combination is significantly more effective than either one alone.

Start your training today at **booksbymario.com/learn**

· · ·

What Comes Next in the AI Advantage Series

This book is the foundation. It covers everything a product manager needs to lead, build, and win in the age of intelligent products.

Book Two in the AI Advantage Series goes deeper on a specific dimension that this book introduces but does not fully develop: **AI product strategy for senior leaders.** It is written for CPOs, VPs of Product, and product directors who are responsible not just for individual AI products but for the organizational capability, portfolio strategy, and competitive positioning that determines whether their company wins or loses in markets where AI is the primary competitive surface. If that is the conversation you need to be having, that is the book that will prepare you for it.

One Last Thing

The introduction to this book opened with a simple framing: two product managers, same company, same day, one earning $245,000 five years later and one earning $123,000. The difference was a decision.

You have made that decision. You picked up this book, read it to the end, and now have the frameworks, the language, and the playbook to be the product manager on the right side of that divide.

What happens from here is not determined by the market, by your company, or by how fast AI continues to evolve. It is determined by what you do with what you now know. The thirty-day plan is the beginning. Everything that follows compounds from there.

The AI advantage is real. It is available. And it starts with the first action you take after closing this page.

— Mario Urbina, May 2026

Training App

Ready to practice what you just read?

The AI Advantage Training App is where the frameworks in this book become fluent, through scenario-based exercises, an AI-powered assessor, and a portfolio builder designed for the AI PM hiring market.

Scan to access your companion training:

Or visit: **booksbymario.com/learn**

Book readers get 30 days free, use code **AIADVANTAGE** *at checkout.*

APPENDICES

Reference material for the practicing AI PM

APPENDIX A
The AI PM Toolkit

The tools listed here are organized by the job they do, not by category or vendor. This list reflects what experienced AI PMs are using as of mid-2026. The AI tooling landscape moves quickly; treat this as a starting point, not a definitive registry.

Discovery and User Research

Dovetail

Use case: Stores, tags, and synthesize qualitative research data. Strong AI-assisted pattern recognition across interview transcripts and usability sessions.

Find it at: *dovetail.com*

Maze

Use case: Runs unmoderated usability tests at scale with built-in AI analysis of task completion and behavioral patterns.

Find it at: *maze.co*

Grain

Use case: Records and transcribes user interviews, generates AI summaries, and allows teams to clip and share key moments.

Find it at: *grain.com*

Product Strategy and Roadmapping

Notion AI

Use case: AI-assisted drafting and synthesis for product principles, strategy documents, and planning artifacts. Strong for the one-page strategy format from Chapter 4.

Find it at: *notion.so*

Productboard

Use case: Connects customer feedback to roadmap items with AI-assisted prioritization. Useful for outcome-based planning.

Find it at: *productboard.com*

Amplitude

Use case: Best-in-class behavioral analytics for product teams. Strong for the user-signal feedback loops described in Chapters 6 and 8.

Find it at: *amplitude.com*

AI Feature Development and Evaluation

LangSmith (LangChain)

Use case: Traces LLM calls, runs evals, and monitors production AI applications. The closest thing to a standard eval infrastructure platform as of 2026.

Find it at: *smith.langchain.com*

Braintrust

Use case: Runs and tracks AI evals across model versions. Useful for the deterministic and probabilistic eval frameworks from Chapter 8.

Find it at: *braintr.ust*

Weights and Biases

Use case: Tracks model training runs, experiments, and eval results. More engineering-oriented but important for PMs to understand.

Find it at: *wandb.ai*

PromptLayer

Use case: Manages, versions, and monitors prompts in production. Useful for the prompt-as-product-decision framework from Chapter 7.

Find it at: *promptlayer.com*

Agentic AI Development

LangGraph

Use case: Framework for building stateful, multi-agent workflows. Useful for understanding the orchestration layer discussed in Chapter 5.

Find it at: *langchain-ai.github.io/langgraph*

Crew AI

Use case: Orchestrates teams of AI agents for complex workflows. Most relevant for the Tier 3 agentic products from Chapter 3.

Find it at: *crewai.com*

Inspect AI

Use case: UK AI Safety Institute's open-source framework for evaluating AI model safety. Relevant for the responsible AI launch checklist from Chapter 9.

Find it at: *inspect.ai-safety-institute.org.uk*

Measurement and Monitoring

Grafana

Use case: Builds the monitoring dashboards for AI product health described in Chapter 8. Widely used for the technical metrics layer.

Find it at: *grafana.com*

Mixpanel

Use case: Strong for the behavioral signal tracking discussed in Chapters 6 and 8. Particularly good for override rate and downstream outcome measurement.

Find it at: *mixpanel.com*

Datadog

Use case: Monitors AI infrastructure performance including latency, error rates, and cost per query. Most relevant for the technical metrics layer.

Find it at: *datadog.com*

Career and Portfolio

Lenny's Newsletter

Use case: Consistently strong signal on what the AI PM hiring market actually values, compensation benchmarks, and emerging role patterns.

Find it at: *lennysnewsletter.com*

LinkedIn

Use case: The primary platform where AI PM portfolio signals are evaluated by hiring teams. The point-of-view post format described in Chapter 10 performs well here.

Find it at: *linkedin.com*

APPENDIX B
The AI Literacy Glossary

Fifty terms every AI PM must know. Definitions are written for product decision-making, not for technical precision. Where a technical definition would be more precise but less useful, the product-relevant version is used. Chapter references indicate where each concept appears in the book.

Agent

An AI system that can take actions, not just generate responses. Agents operate in a perceive-reason-act loop, using tools to interact with external systems on behalf of a user or another agent. (See Chapter 5)

Agentic product

A product whose core value is delivered by AI agents taking consequential actions in the world, as distinct from products that generate responses for humans to act on. (See Chapter 3, 5)

Bias (AI)

Systematic error in AI outputs that affects different user groups, demographic segments, or use cases differently. A product quality problem, a regulatory problem, and an ethical problem simultaneously. (See Chapter 9)

Chain of thought

A prompting technique that instructs a model to reason through a problem step by step before producing an answer. Improves performance on complex reasoning tasks and makes the model's logic more auditable. (See Chapter 7)

Chunking

The process of splitting large documents into smaller segments for use in RAG systems. Chunking strategy significantly affects retrieval quality. *(See Chapter 2)*

Context window

The maximum amount of text an LLM can process in a single inference call. Everything outside the context window is invisible to the model. *(See Chapter 2)*

Data flywheel

A feedback loop where user engagement generates data that improves the AI model, which improves the product, which generates more engagement. The primary source of defensible advantage in AI-native products. *(See Chapter 3)*

Deterministic eval

An evaluation that tests for correctness on questions with clear right and wrong answers: format compliance, required field inclusion, constraint adherence. Should be automated and run on every model change. *(See Chapter 2, 8)*

Embedding

A numerical representation of text that captures semantic meaning, enabling similarity search and retrieval. The foundation of most RAG systems. *(See Chapter 2)*

Eval (evaluation)

The testing framework for AI systems. Determines whether an AI product is performing at the quality standard its users and business require. *(See Chapter 2, 8)*

Fine-tuning

Training a pre-existing model on domain-specific data to improve its performance on a specific task or context. More expensive than prompting or RAG but can produce significantly better results for well-defined tasks. *(See Chapter 2)*

Foundation model

A large AI model trained on broad data that can be adapted for many downstream tasks. GPT-4, Claude, and Gemini are examples. Most AI products are built on top of foundation models rather than training their own. *(See Chapter 2)*

Guardrails

Constraints placed on an AI system's behavior to prevent it from taking actions outside its intended scope. Both a technical implementation and a product design decision. *(See Chapter 5)*

Hallucination

When an LLM generates confident-sounding output that is factually incorrect. A property of how these systems work, not a bug to be fixed, and a design constraint for any AI product that makes factual claims. *(See Chapter 2)*

Human-in-the-loop (HITL)

A design pattern where human judgment is incorporated into an AI workflow at defined decision points. Not a temporary scaffold but a permanent feature of responsible agentic product design. *(See Chapter 5)*

Inference

The process of running a trained AI model to generate an output. Inference cost (latency and compute cost per query) is a key product economics consideration. *(See Chapter 2, 8)*

Knowledge base

The collection of documents, data, and content that a RAG system draws from when generating responses. Quality and freshness of the knowledge base directly determines RAG output quality. *(See Chapter 2)*

Latency

The time between a user's request and the AI system's response. A critical user experience variable that interacts with trust: slow responses reduce perceived reliability. *(See Chapter 8)*

LLM (Large Language Model)

An AI model trained on large amounts of text that generates responses by predicting the most likely next tokens. The foundation of most current AI product features. *(See Chapter 2)*

LLM-as-judge

A pattern where one LLM evaluates the outputs of another LLM against a defined rubric. Enables probabilistic eval at scale without requiring human raters for every sample. *(See Chapter 2, 8)*

Multi-agent system

An architecture where multiple specialized AI agents coordinate to complete complex tasks, each handling a specific domain or function within a larger workflow. *(See Chapter 2, 5)*

Orchestration layer

The component of a multi-agent system that coordinates agents, routes work, manages state, and handles exceptions. A product design surface that reflects the product's principles in edge cases. *(See Chapter 5)*

Overfitting

When a model performs well on its training data but poorly on new, unseen data. A common cause of the gap between eval performance and production performance. *(See Chapter 7)*

Permission creep

The gradual accumulation of more agent permissions than are needed for the current task. A specific agentic product failure mode that creates security and trust risks. *(See Chapter 5)*

Pilot launch

A controlled release of an AI product to a limited user group before full launch. Non-optional for AI products because it catches real-world failure modes that evals systematically miss. *(See Chapter 7)*

Probabilistic eval

An evaluation that requires judgment: is this response helpful, accurate, and appropriate? Cannot be fully automated and often uses LLM-as-judge for scale. *(See Chapter 2, 8)*

Prompt

The input provided to an LLM that shapes its output. A product design artifact, not just a technical configuration. The primary expression of the product's intent. *(See Chapter 7)*

Prompt engineering

The practice of designing and refining prompts to reliably produce the outputs a product requires. A product responsibility, not only an engineering one. *(See Chapter 7)*

RAG (Retrieval-Augmented Generation)

An architecture that retrieves relevant documents from an external knowledge base and passes them to an LLM as context before generating a response. The primary solution to the knowledge gap problem. *(See Chapter 2)*

Reasoning model

An LLM variant that is optimized for multi-step reasoning tasks, typically by generating intermediate reasoning steps before producing a final answer. *(See Chapter 2)*

Retrieval

The process of finding and returning relevant content from a knowledge base in response to a query. The quality of retrieval determines the quality of RAG outputs. *(See Chapter 2)*

RLHF (Reinforcement Learning from Human Feedback)

A training technique that uses human preference judgments to align model behavior with human values and intentions. The primary method by which foundation models are made safe and helpful. *(See Chapter 2)*

Shadow AI proliferation

The emergence of disconnected, independently built AI capabilities across product teams in the same organization. A portfolio-level failure mode that creates redundancy, governance gaps, and compounding technical debt. *(See Chapter 4)*

Silent degradation

A failure mode where AI product quality declines gradually, below the threshold of any single obvious failure, until the cumulative impact is significant. The most dangerous category of AI product quality problem. *(See Chapter 5)*

System prompt

Instructions provided to an LLM before the user's input that establish the model's persona, constraints, and behavioral guidelines. A product design decision with significant user experience implications. *(See Chapter 7)*

Temperature

A parameter that controls the randomness of an LLM's outputs. Higher temperature produces more varied, creative outputs; lower temperature produces more deterministic, consistent outputs. *(See Chapter 2)*

Tier 1 / 2 / 3 AI product

The three-tier classification from Chapter 3: AI-Enhanced (AI adds value to a traditional product), AI-Native (AI is central to the core value proposition), and Agentic (AI agents do consequential work on behalf of users). *(See Chapter 3)*

Token

The unit of text that LLMs process. Approximately three to four characters or three-quarters of a word in English. Token count determines inference cost and context window usage. *(See Chapter 2)*

Tool use (in agents)

The capability of an AI agent to call external functions, APIs, or systems to take actions in the world. The mechanism by which agents do work rather than just generating responses. *(See Chapter 5)*

Trust calibration

The alignment between a user's confidence in an AI system and the system's actual reliability. Miscalibrated trust (over-trust or under-trust) is a product failure, not a user error. *(See Chapter 9)*

Vector database

A database optimized for storing and searching embeddings. The storage layer for most RAG systems. *(See Chapter 2)*

Vibe coding / vibe shipping

Colloquial term for building and shipping AI features based on intuition about what is technically possible rather than validated understanding of user problems and AI limitations. A reliable path to wasted engineering effort. *(See Chapter 7)*

APPENDIX C
The AI Product Metrics Reference

This reference organizes the metrics from Chapter 8 into a single lookup format, organized by the three-layer stack: Technical, Product, and Business. Use it to build your metrics framework before a launch or to audit the completeness of your existing measurement approach.

Layer 1: Technical Metrics

Model latency (p50 / p95 / p99)

Layer: Technical   **Definition:** Time from request to first token (TTFT) and time to complete response. p95 and p99 matter more than average for user experience.

Healthy range / signal: TTFT under 500ms for interactive features. p99 latency within 3x of p50 indicates a stable system.

Error rate

Layer: Technical   **Definition:** Proportion of requests that return an error (model failure, timeout, or invalid output).

Healthy range / signal: Below 0.5% for production AI features. Spikes above 1% warrant immediate investigation.

Hallucination rate

Layer: Technical   **Definition:** Proportion of outputs containing factually incorrect claims, measured through automated fact-checking or human sample review.

Healthy range / signal: Highly context dependent. For factual claim features, define an acceptable threshold before launch and do not ship above it.

Token cost per query

Layer: Technical   **Definition:** Average inference cost per user request. Tracks AI feature unit economics.

Healthy range / signal: Should be modeled before launch. Sustainable unit economics require cost per query to be well below the revenue or value it generates.

Eval suite pass rate

Layer: Technical **Definition:** Proportion of eval cases passing for each eval category (deterministic, probabilistic, adversarial).

Healthy range / signal: Deterministic evals should be at or near 100% before launch. Probabilistic evals require a defined threshold set before the build begins.

Layer 2: Product Metrics

Feature adoption rate

Layer: Product **Definition:** Proportion of eligible users who engage with the AI feature within a defined window after availability.

Healthy range / signal: Highly context-dependent. Low adoption despite awareness is the most important diagnostic signal in the first two weeks post-launch.

Override rate

Layer: Product **Definition:** Proportion of AI suggestions that users modify or reject before acting on them.

Healthy range / signal: Below 15% suggests good alignment between AI output and user intent. Above 30% signals systematic misalignment requiring root cause analysis.

Task completion rate

Layer: Product **Definition:** Proportion of users who complete the target task in sessions where the AI feature was engaged, compared to sessions where it was not.

Healthy range / signal: AI-engaged sessions should show meaningfully higher completion rates. If not, the feature is not creating value in the workflow.

Acceptance rate (generative features)

Layer: Product   **Definition:** Proportion of AI-generated content used without modification by the user.

Healthy range / signal: Above 40% suggests strong relevance. Below 20% suggests the model is frequently missing user intent.

Human intervention rate (agentic features)

Layer: Product   **Definition:** Proportion of agentic workflow instances where a human needed to intervene to correct or redirect the agent.

Healthy range / signal: Define before launch. Higher early rates are expected and acceptable; declining rates over time indicate a learning and improving system.

Irreversible error rate (agentic features)

Layer: Product   **Definition:** Proportion of agent actions that were incorrect and could not be undone.

Healthy range / signal: Should be at or near zero. Any irreversible error above 0.1% in a consequential workflow warrants immediate design review.

Layer 3: Business Metrics

Revenue influenced

Layer: Business   **Definition:** Revenue attributable to AI feature engagement, measured through attribution modeling or controlled experiment.

Healthy range / signal: Establish attribution methodology before launch. Retrospective attribution is significantly less credible with finance and executives.

Cost savings

Layer: Business   **Definition:** Reduction in operational cost attributable to AI automation, measured against a defined baseline.

Healthy range / signal: Define the cost baseline before the AI feature is built. Before-and-after comparisons require stable baselines.

Time saved per user per week

Layer: Business   **Definition:** Reduction in time spent on the target task, measured through session data or user-reported time tracking.

Healthy range / signal: Convert to annual hours to make the business case legible. Ten minutes per user per week is 8+ hours per year per user.

Churn reduction

Layer: Business   **Definition:** Difference in churn rate between users who engage with AI features and users who do not.

Healthy range / signal: AI feature engagement is frequently a leading indicator of retention. Measure at 30, 60, and 90 days post-engagement.

Support volume change

Layer: Business   **Definition:** Change in relevant support ticket volume attributable to AI feature availability.

Healthy range / signal: A reduction signals that the AI is successfully handling requests that previously generated support load.

APPENDIX D
Further Reading and Resources

Organized by the four parts of this book. Each item is included because it goes deeper on a specific dimension that the chapter introduced but could not fully develop. This is a curated list, not a comprehensive bibliography.

Part One: Understand the Shift

Thinking in Systems

Donella Meadows

The best foundation for understanding why AI products behave as systems rather than as features. Essential context for the mental model shifts in Chapters 1 and 3.

The Alignment Problem

Brian Christian

A rigorous, accessible account of the gap between what we build AI systems to do and what they actually do. Deepens the failure mode thinking introduced in Chapter 5.

Prediction Machines

Agrawal, Gans, and Goldfarb

The economics of AI explained through the lens of prediction as a commodity. Essential for the competitive strategy thinking in Chapter 3.

Part Two: Build the Strategy

Inspired (2nd Edition)

Marty Cagan

The product management canon. The adaptive planning framework in Chapter 4 is built in dialogue with Cagan's principles, updated for the AI era.

Good Strategy / Bad Strategy

Richard Rumelt

The clearest articulation of what strategy actually is and what makes it work. Directly applicable to the AI product strategy framework in Chapter 4.

Team Topologies

Skelton and Pais

The organizational design framework most directly applicable to the three AI product org models in Chapter 11. Required reading for CPOs building AI product organizations.

Part Three: Ship Intelligent Products

Continuous Discovery Habits

Teresa Torres

The discovery practice framework that most directly complements the lean AI build cycle in Chapter 7. Particularly strong on connecting user research to product decisions continuously.

Trustworthy Online Controlled Experiments

Kohavi, Tang, and Xu

The rigorous treatment of experimentation that the signal-to-noise framework in Chapter 6 draws from. Essential for teams building serious AI experimentation infrastructure.

The Art of Invisibility

Kevin Mitnick

An accessible treatment of privacy and security thinking that informs the ethical guardrails framework in Chapter 9. More practical than most privacy-focused product resources.

Part Four: Win the Career

Cracking the PM Interview

Gayle Laakmann McDowell and Jackie Bavaro

The foundational PM interview preparation resource, still the best for the product sense dimensions. Supplement with Chapter 10's AI-specific interview preparation.

Radical Candor

Kim Scott

The most practically useful framework for the cross-functional communication and feedback culture that AI PM roles require. Directly applicable to the ML engineer collaboration model in Chapter 7.

Ongoing Resources

The Batch (deeplearning.ai newsletter)

Andrew Ng

Weekly signal on AI research and industry developments. The most reliable single source for staying current on model capability changes that affect product strategy.

Lenny's Newsletter and Podcast

Lenny Rachitsky

The most consistent source of practitioner-level AI PM insight, compensation benchmarks, and role evolution data. Essential for the career positioning work in Chapter 10.

AI Snake Oil (blog and newsletter)

Arvind Narayanan and Sayash Kapoor

The most rigorous public accounting of what AI can and cannot reliably do. Essential reading for calibrating claims in the AI products you build and the strategy you present.

— · —

End of Appendices · The AI Advantage for Product Managers

ACKNOWLEDGEMENTS

Books are announced by one name on the cover. They are built by many.

This one is no exception.

The practitioner community I am close with that made this book possible is too large to name in full, and that is the point. The product managers, engineering leads, data scientists, and product leaders whose real-world decisions, failures, recoveries, and insights shaped the thinking in these pages; you know who you are, even when the details have been composited to protect you. Thank you for doing the work honestly, for sharing what went wrong as readily as what went right.

To the beta readers who helped through early drafts of this book: your feedback was not always comfortable and was always correct. The chapters that read most clearly now are the ones you pushed back on. A book that has not been challenged is a book that has not been finished.

To my colleagues along the way; in corporate offices, on video calls, in conference hallways, and in the kind of late-night conversations that only happen when a product is in trouble and everyone in the room cares enough to stay: you taught me more than any framework I have read.

To the voices I follow online and the people I have met in communities, at events, and in the quiet corners of the internet where serious practitioners gather to think out loud: you have shaped how I see this profession without always knowing it. The generosity of people who share what they are learning, in public, for the benefit of strangers, is one of the things I admire most about the technology community.

To my friends in tech, the ones who have been there through every version of this journey, who have pushed me to go further when I thought I had gone far enough, who have called out the gaps in my thinking with the particular directness that only comes from people who genuinely want you to be better: I would not have written this without you.

And to every product manager reading this who is trying to figure out which side of the AI divide to stand on: this book was written for you.

— Mario Urbina, Connecticut, May 2026

— · —

End of The Book · The AI Advantage for Product Managers